## ALSO BY NEIL STRAUSS

*Rules of the Game*

*The Game: Penetrating the Secret Society of Pickup Artists*

*The Dirt*
with Mötley Crüe

*How to Make Love Like a Porn Star*
with Jenna Jameson

*The Long Hard Road Out of Hell*
with Marilyn Manson

*Don't Try This at Home*
with Dave Navarro

*How to Make Money Like a Porn Star*
with Bernard Chang

# EMERGENCY

THIS BOOK WILL SAVE YOUR LIFE

# NEIL STRAUSS

HARPER

NEW YORK • LONDON • TORONTO • SYDNEY

## HARPER

To protect the innocent, the names and identifying details of a small number of individuals have been changed and two characters are composites.

Photographic Credits and Permissions
p. 21 As published in the *Chicago Sun-Times*. Reprinted with permission.
p. 55 (bottom) Prestel Publishing Ltd.
p. 58 Natalie Behring
p. 87 Keystone/Hutton Archive/Getty Images
p. 131 Book cover reprinted with the permission of the Sovereign Society.
p. 150 Tomas Skala
p. 293 and 297 Kristine Harlan
p. 347 Courtesy of Sport Copter, Inc.
All other photographs taken by the author.

The activities in this book involve the use of tools, resources, and materials that can be dangerous and require training, supervision, and practice in order to be used safely. Your safety is your responsibility, including the proper use of equipment and safety gear and determining whether you have the adequate skill and experience required for an activity. Please see contract for further details.

FIRST EDITION

www.neilstrauss.com

*Designed by Todd Gallopo and Jaime Putorti*
*Illustrations by Bernard Chang*

Library of Congress Cataloging-in-Publication Data is available upon request.

ISBN 978-0-06-089877-9

09   10   11   12   13   DIX/RRD   10   9   8   7   6   5   4   3   2   1

# CONTRACT OF LIABILITY

I,_____, being of sound mind and body, have hereby chosen of my own accord to read the book EMERGENCY. I believe that, on the present date of _____ / _____ / _____, my chances of surviving a world crisis, systemic shutdown, and mass hysteria are _____ percent. I hereby resolve that after reading this book and committing to my own betterment and self-sufficiency, my chances of survival will become _____ percent. Therefore, I understand that the information in this book has been carefully researched, and all efforts have been made to ensure accuracy. And, moreover, I acknowledge that the author and the publisher assume no responsibility for any injuries suffered or damages or losses incurred during or as a result of following this information. I will carefully study, independently research, and clearly understand all information before taking any action based on this book. I assume full responsibility for the consequences of my own actions. I will not play with guns, knives, fire, wild plants, wild animals, wild police officers, foreign governments, the IRS, or myself. And, if I do not survive the apocalypse, I will not hold HarperCollins responsible. By means of turning to another page, closing this book, or moving my eyes away from this line of type, I agree to make this contract irrevocably binding.

I hereby name _____ who resides at _____ as my next of kin and beneficiary of this book.

TO DONALD BOOTH, WHO DIED
WHEN A BLOCK OF ICE FELL FROM
A BUILDING AND HIT HIS HEAD.
AND TO ALL THOSE WHO NEVER
SAW IT COMING.

MEMENTO MORI...

THERE IS NO CRIME
THAT A MAN WILL NOT COMMIT
IN ORDER TO SAVE HIMSELF.

—Tadeusz Borowski, "The January Offensive"

**PART ONE**

# ORIENTATION

Notice the strong walls of our city . . .
Now examine the inner walls of our city.
Examine the fine brickwork.
These walls, too, surpass all others!
No human being, not even a king,
will ever be able to construct more impressive walls.

—*Gilgamesh*, Tablet I, 2100 B.C.

# PROLOGUE

*Ring. Ring.*

The time was 7:40 A.M. I reached for the phone.

"Do you have your axe?" came the voice on the other end. It was Mad Dog.

"Yes."

"Is your axe sharp?"

"No, but I can sharpen it while you're driving here."

"How about your knife?"

"Got it."

"Everything needs to be nice and sharp."

Fuck, I'm supposed to kill a goat today. And I couldn't even kill the fly in my room last night. Really. Sadly. I just put a drinking glass over it, covered the opening with a saucer, then set it free outside. I'm a victim of my own empathy. I wouldn't be too happy if someone squished me flat, so it seems cruel to do the same to another living thing.

Fifteen minutes later, Mad Dog pulled up in a weathered blue Dodge Ram 3500 truck with skull-and-crossbones floor mats and a lone bumper sticker depicting a gun sight next to the words THIS IS MY PEACE SYMBOL.

The goat peered curiously at me from a beige dog cage in the

3

back of the truck. It was much cuter than I'd expected. It had a wide smile, silky white fur, and a gentle disposition. I began to feel sick.

Symptoms: dizziness, nausea, shortness of breath.

I turned away. I didn't want to pet it, befriend it, name it, or grow attached to it in any way. If I did, there was no way I'd be able to go through with this.

My girlfriend Katie, whom I'd brought along for moral support, felt the same way. "Oh my God—it just *baa*'d at me," she squealed in delight and horror. "I can't look. I'll fall in love."

So much for moral support.

"Is this wrong?" I asked Mad Dog as we drove into the forest in grim silence. "I need a moral justification for doing this."

"This is the circle of life," he answered coldly, without sympathy. He was thin, with ropy muscles, a receding hairline, piercing blue eyes, and a brown handlebar mustache. His hat was emblazoned with the Revolutionary War slogan "Don't tread on me," and he wore a sleeveless T-shirt advertising his handmade knives.

"Every steak you bought at Safeway started out looking like this," he continued. "If you need a rationalization, you're hungry and you need to eat today. And if you want to eat, something has to die." Then he leaned forward, flipped on his stereo, and blasted AC/DC's "Kicked in the Teeth."

Unlike me, Mad Dog was a real man. He could chop wood, make fire, forge weapons, kill his own food, and defend himself with his bare hands. In other words, he could survive on his own—without Con Edison, without AT&T, without Exxon, without McDonald's, without Wal-Mart, without two and a half centuries of American civilization and industry.

And that's exactly why I was with him right now, crossing a moral boundary from which there was no return.

"Help me look for a good hanging tree," Mad Dog ordered

as he stopped at a clearing deep in the woods and turned off the engine.

Every moment, this felt more and more like a Mafia execution. In the distance, I saw a deer bound across a clearing and disappear into the forest. It was such a strong, beautiful, graceful animal. I didn't think I could ever shoot one.

Unless Mad Dog told me to.

After finding the tree and throwing pigging string over a branch, we returned to the truck and stood at the rear bumper next to the goat cage. "This is your protein source," Mad Dog began his lecture. "Right along its neck is its carotid artery. You're going to straddle the goat, push your knife through from one side to the other, and cut out the throat. Then we're going to hang it, skin it, and butcher it."

Symptoms: dizziness, nausea, shortness of breath, self-disgust, guilt.

He let the goat out of the cage and put a leash around its neck. It walked up to me and nuzzled its head against my leg. Then it stepped away and peed and shat on the ground.

"The more waste it passes now," Mad Dog said, "the better."

This was when reality set in. I felt, in that moment, like I was going to hell. The goat was able to handle a leash, and it waited until it was out of the cage to relieve itself. It was practically domesticated.

I didn't have to kill it. I could always ask Mad Dog if I could just keep it as a pet.

"Don't anthropomorphize your prey," Mad Dog barked when I confided this to him. "Most animals won't piss and shit where they lay down."

"I've been trying not to get attached," I told him. "That's why I haven't given it a name."

"I have," Katie blurted. "I named it Bettie. B-E-T-T-I-E."

"When did you do that?"

"When she fluttered her little eyes at me."

That was the last thing I needed to hear.

Symptoms: everything, nothing, complete and total panic.

I wasn't sure I could go through with this.

I was wearing an olive baseball cap, a matching army shirt, khaki cargo pants, and a gun belt with a Springfield Armory XD nine-millimeter on one side and a three-inch RAT knife on the other. This wasn't me. Until a month ago, I'd rarely even worn cargo pants or baseball caps, let alone guns or knives.

Why, I asked myself, was I about to do this?

Because I wanted to survive. This is what people did for protein before there were farms and slaughterhouses and packing plants and refrigerated trucks and interstate highways and grocery stores and credit cards.

I never thought the day would come when I'd have to make a backup plan.

# A BRIEF CONFESSION

I've begun to look at the world through apocalypse eyes.

It usually begins in airports. That's when I get the first portent of doom. I imagine explosions, sirens, walls blown apart, bodies ripped from life.

Then, as I gaze out of the taxi window after arriving in a new city, I see people bustling around on their daily routine, endless rows of office buildings and tenements teeming with activity, thousands of automobiles rushing somewhere important. And it all seems so solid, so permanent, so unmovable, so absolutely necessary.

But all it would take is one war, one riot, one dirty bomb, one natural disaster, one marauding army, one economic catastrophe, one vial containing one virus to bring it all smashing down. We've seen it happen in Hiroshima. In Dresden. In Bosnia. In Rwanda. In Baghdad. In Halabja. In New Orleans.

Our society, which seems so sturdily built out of concrete and custom, is just a temporary resting place, a hotel our civilization checked into a couple hundred years ago and must one day check out of. It's an inevitability tourists can't help but realize when visiting Mayan ruins, Egyptian ruins, Roman ruins. How long will it be before someone is visiting American ruins?

That's how the world looks through apocalypse eyes. You start filling in the blanks between a thriving city and a devastated one. You imagine how it could happen, what it would look like, and whether you and the people you love could escape.

Of course I don't want it to happen. Hopefully, it will never happen. But for the first time in my life, I feel there's a possibility it will. And that's enough to motivate me. To motivate me to save myself and my loved ones while there's still time.

I don't want to be hiding in cellars, fighting old women for a scrap of bread, taking forced marches at gunpoint, dying of cholera in refugee camps, or anything else I've read about in history books. I want to be writing those history books on a beach far away from the mess that self-serving politicians, crooked CEOs, and committed madmen are making of the Western world.

I want to be the one who gets away. The winner of the survival lottery.

I didn't always think like this. But then again, I was naive. I belong to the American generation that believed it was beyond history. Until this millennium, nothing bad had happened to us like it had to every generation before. Those who came of age in the first twenty years of the century had World War I. The next twenty years were marked by the Great Depression. The following twenty years began with World War II. The next generation inherited Vietnam.

And then, from 1980 to the close of the century—nothing. Or at least no war, no national catastrophe, no defining event powerful enough to pull us outside our self-centered, solipsistic world, outside our preoccupation with ourselves and our financial and emotional well-being, outside our comfort zone.

Of course, society wasn't perfect, but to many Americans, it felt like we were just a cure for AIDS, a solution to the drug prob-

lem, and an effective campaign against urban gang violence away from getting as close as possible.

But then, swiftly and without warning, it happened.

History happened to us.

Terrorist attacks. Domestic crackdowns. Flooded cities. Bank failures. Economic collapse.

I can't tell you the exact date along the way I lost faith in the system, because for me there were five of them, each chronicled in the section that follows. And over the course of this gradual awakening—which perhaps coincidentally, perhaps not, covered the span of the Bush administration—I decided to equip myself with the tools necessary to survive whatever politics and history threw at me next.

By the time the Obama administration stepped in with a message of hope and change, it was too late to undo the damage. Because I now know that, even in America, anything can happen.

Preparing myself for hard times has been an incredibly challenging task, because some people were born tough. I wasn't. My parents live on the forty-second floor of a seventy-two-story building in Chicago. They didn't camp, hunt, farm, cook, or even fix things themselves.

As for learning skills after leaving home, I spent most of my adult life as a music writer for the *New York Times,* so I could tell you anything you wanted to know about rock and hip-hop, but nothing about growing food or building fires or defending yourself. In fact, I'd never even been in a fight in my life, though I had been mugged twice.

In short, if the system ever did break down, the only useful skill I really had was the ability to write about it. Perhaps, at best, I could talk someone with practical knowledge into helping me out. Or maybe they'd just mug me.

But that wouldn't happen anymore. Today I can draw a holstered pistol in 1.5 seconds, aim at a target seven yards away, and shoot it twice in the heart. I can start a fire by rubbing two pieces of wood together. I can identify seven hundred types of footprints when tracking animals and humans. I can survive in the wild with nothing but a knife and the clothes on my back. I can find water in the desert, extract drinkable fluids from the ocean, deliver a baby, fly a plane, pick locks, hot-wire cars, build homes, set traps, evade bounty hunters, suture a bullet wound, kill a man with my bare hands, and escape across the border with documents identifying me as the citizen of a small island republic.

When the shit hits the fan, you're going to want to find me. And you'll want to be doing whatever I'm doing. Because I've learned from the best.

You can call me crazy if you want.

Or you can listen to the story of the eight years it took to open my eyes, realize my country can't protect me, and do something about it.

It just may save your life.

# HOW TO TURN A CREDIT CARD INTO A KNIFE

| 1. SELECT A STIFF CARD, SUCH AS A CREDIT CARD OR HOTEL ROOM KEY CARD. | 2. TAKE A SINGLE-EDGED DISPOSABLE RAZOR, SUCH AS A PLASTIC BIC. | 3. BREAK THE PLASTIC HEAD APART AND REMOVE THE BLADE. | 4. SUPERGLUE THE BLADE TO THE CORNER OF THE CARD AT AN ANGLE. IT SHOULD PROTRUDE ABOVE THE EDGE OF THE CARD ABOUT 1/16TH OF AN INCH. |
| --- | --- | --- | --- |

5. COVER THE LOWER PART OF THE BLADE WITH SCOTCH TAPE TO KEEP IT FROM SNAGGING WHEN MAKING DEEP CUTS.

6. KEEP THE CARD BLADE-DOWN BEHIND YOUR DRIVER'S LICENSE (BECAUSE THE BLADE PARTS THE FLESH AHEAD OF THE CARD, YOU CAN MAKE DEEP SLICES).

VESA
42K6 5467 55

7. THIS WAY, IF SOMEONE WITH MURDEROUS INTENT ASKS YOU FOR YOUR ID OR BANK CARD, YOU CAN WHIP IT OUT AND SLICE THEIR NECK.

8. BECAUSE OF THE SMALL BLADE, IT CAN BE CARRIED SAFELY THROUGH MOST METAL DETECTORS.

TO BE CONTINUED...

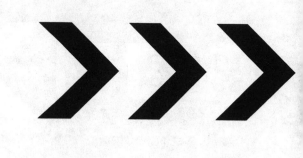

## PART TWO

# FIVE STEPS

Only the gods can dwell forever with the Sun.
As for the human beings, their days are numbered.
And it is no more than blustery weather,
No matter what they try to achieve.

—*Gilgamesh*, Tablet II, 2100 B.C.

# STEP 1: DECEMBER 31, 1999

## LESSON 1

# BIRTHDAY CLOWNS TO AVOID

**Y**ou need to pick a group that won't kill you."

The voice on the phone was that of Jo Thomas. A fellow *New York Times* reporter, she was on the cult and terrorism beat. She'd interviewed Timothy McVeigh after the Oklahoma City bombing, covered the Sinn Fein in Northern Ireland at the height of their reign of terror, and investigated the aftermath of David Koresh and his bloody last stand against the FBI in Waco.

I had just volunteered to spend New Year's Eve 1999 with a death cult. It seemed like a good idea at the time. But, just to be safe, I'd called Jo for advice.

The newspaper was sending reporters to different locations to prepare a package of features on the millennial moment. And I wanted to take part in it. I envisioned a group of middle-aged men and women on a remote hillside, clasping hands and awaiting the apocalypse. And I wanted to see the look on their faces when the world didn't end at the stroke of midnight. I wanted to hear how they would rationalize it afterward.

Back then, I had no idea that I'd ever feel unsafe in America or be preparing for disaster myself. We seemed to stand monolithic and invulnerable at the center of the political, cultural, and moral universe, unchallenged as the world's lone superpower. For all the headlines screaming doomsday and worldwide computer shutdown, no sane person really believed life was going to come to an end just because a calendar year was changing. We'd survived the last millennium well enough.

But there were some very panicked people out there who truly didn't think we'd make it to January 1. And those people, Jo warned, were not just likable kooks.

"I don't think anyone in New York knows how scary these groups are," she explained. "A lot of them are nuts who stockpile guns. And most of them consider the media the enemy . . . especially the *New York Times.*"

She then gave concrete examples of just how dangerous these groups could be. One antigovernment militia group in Sacramento had just been busted for planning to incinerate two twelve-million-gallon propane tanks to start a revolution for the New Year. And a second group, calling itself the Southeastern States Alliance, had been caught three days earlier trying to blow up energy plants in Florida and Georgia.

"That's crazy," I thanked her for the advice. "I'll definitely be careful with this."

That didn't satisfy her. "I don't know how old you are," she warned before hanging up, "but however old you are, you're not ready to leave this world."

Death isn't something we're born afraid of. It's something we learn to fear. According to studies, children have little conception of death up to age five. From five to eight, they have a vague understanding of the finality of death. Only at nine do they begin to understand that death is something that one day may happen to them.

My awakening came at the age of nine, thanks to the copy of the *Chicago Sun-Times* that my parents left on the kitchen table every day. One morning, this caught my eye:

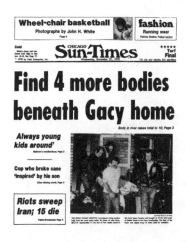

I sat down and read the story. Dozens of bodies of young boys, many of them close to my age, had been found buried in a basement and yard in the northwest section of Chicago, my hometown. A birthday clown named John Wayne Gacy had tortured, molested, and killed them. From that day forward, I realized I was no longer the master of my own safety. It wasn't just climbing trees and running with scissors that could harm me—it was other people.

Before making my decision about the millennium, I called a friend of Jo's at the Southern Poverty Law Center, which tracks cults and hate groups, and asked him to recommend a few relatively safe sects to celebrate with.

"There's a very anti-Semitic fascist group called the Society of St. Pius X in Kansas you might want to look into," suggested Mike Reynolds, one of the center's militia task force investigators. "They're probably not going to do anything to you."

"Probably?"

"Well, there's also William Cooper, who heads a militia group in Arizona. He's training them to go to war after New Year, when Satan is supposed to appear. Or you can try Tom Chittum, who's looking to start a race riot, which he calls Civil War II. Maybe that would be too dark for you. Then there are the Black Hebrew Israelites in Chicago . . ."

Clearly Mike didn't care if I survived the New Year.

Despite the Oklahoma City bombing five years earlier, I had no idea there were so many networks actively trying to destroy America from within. Where reading about John Wayne Gacy had woken me up to the danger lone madmen posed to my safety, talking with Reynolds opened my eyes to the existence of organized groups of them. So in light of this information, I decided to narrow my search to more friendly, unarmed, cuddly doomsday groups.

The next day, I began sending solicitous e-mails to various doomsayers and survivalists, asking if I could spend a few hours with them as the year changed over. I promised to bring my own food, water, and emergency supplies, hoping that somehow this would convince them I was a believer.

I soon discovered that one of the difficulties in writing about people who think the world is going to end is that they instantly know you don't believe them. Because if you did, you'd know there wouldn't be anyone left the next day to read your article.

The first person I contacted was Thomas Chase, a writer and theorist who predicted that the millennium bug would cause a massive electrical crash, triggering a worldwide depression and the coming of the Antichrist. I wondered what kind of sacred and meaningful ritual he'd be performing to prepare for the terror of the apocalypse.

"I plan on going to Boston's First Night celebration with my wife, Peg," he responded. "I've usually gone every year for the last few years."

What about preparing for the End Times? "I did stock up on some extra water," he offered.

The prospect of spending New Year's Eve on the Charles River, then going back to Chase's house to drink water with his wife wasn't exactly Pulitzer Prize material. So I decided to call Jack Van Impe, a televangelist who'd been preaching the apocalypse since before it was trendy. He'd been warning viewers regularly to prepare for the coming devastation of Y2K.

But when I asked Rev. John R. Lang, the executive director of his ministry, what Van Impe was doing to prepare, he told me their leader planned to "ring in the New Year" with Mrs. Van Impe and family at home watching television.

After four more calls with similar results, the whole Y2K doomsday thing began to look like a big hoax. Perhaps it was just simple economics: with tabloid readers and journalists (including me) clamoring for people who thought the world was going to end on January 1, 2000—after all, something significant should happen to commemorate such a lovely round number— scores of attention-hungry people arose to fulfill the demand. This was before the reality TV boom. There were fewer routes to national humiliation back then.

So, in an act of desperation, I decided to ignore Jo Thomas's advice and contact the most dangerous group on the list.

## LESSON 2

# TIPS ON DEATH CULT ETIQUETTE

Despite having a rockabilly singer for a leader, the House of Yahweh was not a house to fuck with.

An apocalyptic cult run by Yisrayl Hawkins (known as Buffalo Bill in his rockabilly days), the House of Yahweh had received a lot of heat after Waco because of its similarities to the Branch Davidians. As a result, it was being watched closely by the FBI's antiterrorist Y2K task force.

Because its members are secretive to the point of paranoia—even posting armed guards around their forty-four-acre Texas compound—I dialed their bunker with some trepidation.

"Hello, how may I help you?" a woman's melodious voice answered.

I was taken aback. She seemed friendly. "Is this the House of Yahweh?" I asked.

"It is," she said warmly, professionally. "What extension may I transfer you to?"

She sounded like a law firm secretary. The only difference was, at this law firm it would be impossible to look people up by their last names: they had all changed their family name to Hawkins in honor of their leader.

"Do you have, like, a publicist?" I stammered. "Or maybe someone who sort of deals with the press?"

"One moment," she sang into my ear. "Please hold for Shaul Hawkins."

The great thing about real life is that it will always surprise you. Nothing ever turns out the way you expect. I suppose that's why I write nonfiction. If this were a movie, the organization would already have traced my number, bugged my phone, and kidnapped my brother. Instead I was being transferred to the publicist and media relations executive for a death cult.

"I'm doing a story for the *New York Times*," I told Shaul when he answered. "They're sending reporters to different places to, um, ring in the New Year. And I wanted to see what you were doing."

I'm the worst reporter, because I get nervous every time I talk to someone. Instead of sounding like a sharp, tough-minded journalist fighting for truth, I sounded like I was asking him out on a date.

"Nothing out of the ordinary," he responded.

I wasn't surprised anymore.

He went on to explain that the group doesn't believe in the year 2000. They believe in the Torah and follow the Hebrew calendar, in which the year is 5760. The only thing happening in the compound on New Year's Eve, he said, would be their normal Friday Sabbath celebration.

He sounded like a nice guy, normal, someone I could hang out with. But then he continued. They always seem to do that.

"We do encourage people to store food, but not for the year 2000. After the Y2K fears don't pan out, everyone is going to think the world is okay. But in the book of Isaiah, it talks about the earth being burned and no one left. The only way this is going to happen is with nuclear weapons."

All right, now I was getting somewhere. At least they thought the world was going to end—maybe not on December 31, but sometime.

"It's going to happen soon, and it's going to be over the seven-year agreement that took place on the White House lawn between [Yitzhak] Rabin and [Yasser] Arafat," he went on. "In the news, Russia and China met and told the U.S. to stick its nose out of their business. It's a very small incident, but it could also lead to major upheavals and terrorist actions. Even Clinton was warning that we're facing biological and chemical warfare and mentioned giving out gas masks."

He paused for effect, then concluded: "There isn't going to be any warning."

"That makes sense," I replied. Those words actually came out of my mouth. I'm a very empathic person. I tend to see a person's point of view easily, even if he's criminally insane. But it does seem to be a good way to make new friends, because moments later Shaul was inviting me to join the group on a pilgrimage to Israel.

"I'll see if the paper will let me," I said.

Why do I say these things?

I hung up and returned to my research. I had my heart set on a sleepover with a doomsday group. After all, New Year's Eve had always been anticlimactic. The year before I'd been at a party in a studio apartment where I only knew one person. When midnight came, I just stood there like an idiot, weakly mouthing "Happy New Year!" to anyone who accidentally made eye contact with me. So the year 2000 promised to be extra anticlimactic—unless I could find someone who wasn't just paying lip service to the apocalypse.

The solution came in the form of a follower of Gary North, a Christian Reconstructionist and the Typhoid Mary of Y2K paranoia. Since 1998, after what he claimed were four thousand

hours of research, he'd been warning that when the clock struck midnight, power plants, which run on preprogrammed computer chips supposedly unable to handle the changeover from 99 to 00, would shut down. This, he predicted, would lead to a domino effect of disasters and riots that could result in two billion deaths.

Though North and one of his predecessors, the survivalist pioneer Kurt Saxon, weren't speaking to the media, I learned that a group of their followers had built a self-sufficient community called Prayer Lake in the hills outside Huntsville, Arkansas.

They believed that with faith alone they would weather the coming devastation. They also stockpiled some food, water, and emergency supplies in case they ran out of faith. Unlike other survivalists, they didn't have guns or artillery. Rather than training to fight looters, they built additional homes and saved extra food to give potential robbers as a peace offering.

Thanks to the miracle of directory assistance, I found a phone number for Bob Rutz, who had come up with the idea for Prayer Lake. I imagined sitting in Rutz's new house with his family, praying and waiting to see what happened at midnight. But of course he had other plans.

"There's a countywide event at the local skating rink," he told me. "We'll be there skating, praying, and eating."

Though Rutz was reluctant to talk to the press about Y2K for fear of being labeled a "crazy," he spent the next half hour speaking with me about it anyway. "I believe it's going to be very bad, but I'm not going to be very worried about it," he said. "All I can do is have faith in God."

President Clinton, Rutz believed, was planning to take complete control of the country by using the Y2K panic as an excuse to enact martial law.

"So many simultaneous things are fixing to happen," he continued. "The Chinese have an agenda, the Iraqis want to wipe us off the map, the Russians have a use-it-or-lose-it mentality, and the Muslim terrorists want to destroy us. The other line of evidence I'm looking at is the amount of oil getting to your gas tank. I used to be an engineer for Fluor in the Persian Gulf, and I know those legacy computer systems. If we had three more years and a couple million dollars more, we still couldn't get ready. All you can do is get to know the Lord better."

I thought Rutz might be the one. Sure, he was going to a skating rink in Huntsville, but at least I'd get some interesting conspiracy theory, a handful of Scripture quotations, and a good headline like "The Last Skate."

"If God tells you that you should be with us, then you may," he finally said. "Come on over to the rink at six or seven in the evening. We'll be together—whatever happens."

He recommended a place to stay in town called the Faubus Motel, so I called and asked if they had any rooms available on December 31. "When the rollover comes," the owner warned in a slow Southern drawl, "we're not responsible if the utilities go out. You can't get your money back or nothing like that."

I asked if I should bring anything in case that happened.

"Well," he replied, taking his time with each syllable, "I served in the Gulf War. And there's something my commanding officer told me that I will now tell you: 'A good soldier is always prepared.'"

The next day, I set the afternoon aside to prepare. I didn't know of any survival supply stores in the neighborhood, so I went to a corner deli and bought a bag of beef jerky, a flashlight, and a bottle of water. Realistically, I doubted anything was going to happen, and even if it did, I knew Rutz had stocked up on extra provisions for looters. So all I had to do was start looting, and the treasure would be mine.

However, there was one last item I knew I'd need to bolster my credibility with Rutz. So I went to a bookstore and bought a Bible.

I still had two weeks until the New Year. And, despite my skepticism, the closer it approached, the more my heart tightened.

In the late nineties, Western civilization appeared to have advanced beyond religious wars, beyond genocide, beyond imperialism, beyond borders. The Cold War had ended, the euro had been introduced, America was experiencing the longest economic boom in its history, and the Internet and the mobile phone were turning the world into a neighborhood with few secrets.

The political philosopher Francis Fukuyama captured the spirit of the time best in his 1989 essay "The End of History." "What we are witnessing is not just the end of the Cold War, or a passing of a particular period of postwar history, but the end of history as such," he wrote. "That is, the end point of mankind's ideological evolution and the universalization of Western liberal democracy as the final form of human government."

We are a plot-oriented species, in perpetual need of satisfying conclusions to our stories. Even the Bible skips ahead to offer the Book of Revelations as closure to Genesis. And so, one couldn't help but wonder, if we truly were at the end of history, then what was going to happen next in the story of our civilization, our species, our planet? How would we end?

As Y2K drew closer, I began to get obsessed with those questions. When I talked to friends, all I could discuss was the millennium and what was going to occur. In my heart I knew everything would be fine, but in my head I imagined the worst. All the extremists I'd talked to were getting to me.

I didn't know it at the time, but these were the first tremors of an earthquake that would eventually awaken the survivalist lying dormant in me—as well as in some of the most successful businessmen in the country, who would become unlikely allies in my obsessive quest. Perhaps we make fun of those we're most scared of becoming.

Soon, it became impossible for me to think beyond December 31, to make any sort of plan after then, or even to write anything due after that date. I wanted to wait it out, skate with the survivalists, and hold my breath until the calendar year changed. Then I could exhale.

But then something unexpected happened: I received an invitation to the White House.

## LESSON 3

# THE USE OF PHILOSOPHY AS AN EXCUSE TO CANCEL PLANS

I'd mainly known Bianca Gilchrist through her answering machine messages. They were short, shrill, and sharp. She was strictly business-class. Her world revolved around her position and responsibilities as a publicist in the country music industry. So she'd call offering me an interview with Johnny Cash or Dolly Parton, and usually I'd take her up on those offers, because I liked the people she worked with.

I never would have guessed that each yes was to her a sign not just of a successful work call and a faithful execution of her responsibility to a client, but of a deepening connection to me, until, without even meeting me, she began to grow attracted to this black-ink byline in the *New York Times*. Unfortunately, I had a rule never to sleep with anyone in the music business—not out of any personal morality, but because I'd failed in all my previous attempts and it was just embarrassing.

So three days before the millennium, Bianca called.

"D'ya wanna go to the White House with Trisha and me?"

By Trisha, she meant country singer Trisha Yearwood, whose record label, MCA, she'd recently been hired to work at. "When would that be, exactly?" I asked.

"For that Millennium Concert at the Lincoln Memorial. There's a party at the White House after and all." She always talked like she was chewing gum between words. "They're flying us there in a private jet. Ya don't need to write about it. Just come as Trisha's guest. It'll be fun."

"Shit, I'm supposed to go ice-skating with some guys who think the world's going to end. Give me a day to figure things out and I'll get right back to you."

"Okay, but ya better hurry. I can invite anyone I want, y'know."

And so began twenty-four hours of mental agony, because each option appealed to a different part of my personality. My philosophy on life is that until I see it proven otherwise, I only have one to live. Even if there's a heaven or there's reincarnation or our energy exists forever, there's no telling whether our memories or our conscious mind will survive the trip. So I want to pack as much into this lifetime as I can. As big as the world is, I want to see it, do it, learn it, experience it all—as long as it doesn't hurt me too much or others at all. And since I would only get to experience one millennial New Year's Eve in this lifetime, I needed to decide whether I wanted to spend it in the seat of American power in Washington, DC or the seat of American paranoia in Huntsville.

Either way, if anything did happen, I couldn't imagine two safer places to be.

In the end, I chose power over paranoia. The only complication was that, in the meantime, I'd realized Bianca wasn't inviting me because she wanted press coverage. She wanted me. Accepting her invitation meant making a prostitute of myself.

And I didn't have a problem doing that.

I rented a tuxedo and filled my suitcase with the provisions I'd originally bought for Huntsville: flashlight, granola bars, beef

jerky, and, of course, the Holy Bible. For some reason I also packed a pair of binoculars. A good soldier is always prepared.

On the morning of December 30, Bianca picked me up and drove me to a private airstrip. She was short, heavyset, and slightly freckled, with stringy blond hair. Like many in the industry, she had a brittleness to her, as if in order to succeed in a man's world she had to sacrifice some of the softness and submission that serve as honey to men on a date but as weakness in an office.

We parked on the tarmac alongside the plane and walked onboard with our bags to find Yearwood in a foul mood because her boyfriend, Garth Brooks, wasn't coming. When we arrived in DC, a limousine picked us up at the airport, dropped our bags at the Madison Hotel, and then took us to rehearsal at the Lincoln Memorial, where an enormous stage had been erected.

The main event was still a day away, and black-suited Secret Service men were already posted everywhere. Backstage, agents mingled with Luther Vandross, Tom Jones, Will Smith, Quincy Jones, Slash, and other celebrities. Several hundred feet above, sharpshooters perched atop the memorial aimed their weapons at us.

Watching the elaborate security procedures, it seemed like there was little difference between the fringe lunatics and the men in power. Each was fueled by paranoia about the other. While the radicals bunkered up in fear of the government, the government bunkered up in fear of the radicals.

"We're wondering right now if the Y2K bug has already hit," a White House employee told Yearwood. "The security clearance cards we use weren't working today. And I heard the computers at one of the newspapers here crashed, and they had to lay it out by hand."

"I think that's a separate issue," a voice interrupted. I looked over to see John McCain, who was running at the time for the Republican nomination in the 2000 presidential election. He was trailed by an eager, just-out-of-college assistant. Easily ingratiating himself into the conversation, McCain seemed more cavalier about the millennium than any of us. However, he had played his own part in the panic, introducing a bill to restrict lawsuits against technology companies for Y2K problems.

The longer the celebrities and politicians talked, the more they admitted their fears: of losing phone reception, of being trapped in DC, of being cut off from food and heat. Because they'd made no preparations, these mainstream role models were actually more nervous about a millennial apocalypse than the sham prophets and cult members, who had at least accepted their potential fate.

Though few people know it, America was founded with the apocalypse in mind. Christopher Columbus wasn't just searching for gold or spices or a new trade route to Asia when he discovered the continent. He believed the world was about to end, and his mission was to save as many souls as he could before the clock ran out.

In his letters to the king and queen of Spain soliciting funds for his next and final expedition, Columbus wrote that "only 155 years remain of the 7,000 years in which . . . the world must come to an end." According to his interpretation of biblical prophecy, his voyages to the New World were the first step toward the liberation of the holy land of Jerusalem from Muslim domination—which would be followed, he wrote, by "the end of the religion of Mohammad and the coming of the Antichrist."

So from the day it was discovered, North America was a portent of doom—a catalyst for a coming apocalyptic war pitting Christians against Muslims. Two centuries later, John Winthrop led the Puritans to America not just for religious freedom but

because he was running from a supposed apocalypse. In his case, he believed God was going to destroy England.

Thus, our founders were actually cut from the same cloth as zealots like Yisrayl Hawkins and Bob Rutz. Even more disturbing, this zealotry still dominates the country today. According to a recent CNN poll, 57 percent of Christians in America believe that the prophecies in the Book of Revelations will literally come true—and one in five of those believes it will happen in their lifetime.

As I thought about America's unceasing obsession with fire and brimstone, I wandered out of the backstage area, awkward around the politicians, stars, and sycophants, to watch the stage crew. Several workers were in the process of hurling curses at an artificial sun that refused to illuminate. They'd evidently spent three weeks and $3 million in taxpayer money building it. On-stage, Will Smith rehearsed the song he'd written to exploit the millennium:

> *What's gonna happen? Don't nobody know.*
> *We'll see when the clock gets to 12-0-0.*
> *Chaos, the cops gonna block the street.*
> *Man, who the hell cares? Just don't stop the beat.*

The Secret Service, however, wanted to stop the beat. When I returned backstage, Yearwood was in a heated argument with several dark-suited men. She had planned to open the show with a snippet of Bob Dylan's "Blowin' in the Wind," but they told her the song was inappropriate and refused to explain why. Later, I asked one of the show's producers about it.

"They thought the lyrics were a bomb reference," he explained.

Only in Washington would a song that had been the anthem

of the peace movement for thirty years be interpreted as a terrorist plot.

The closer we moved toward the millennial moment, the more ridiculous people seemed to be getting. But this was politics after all. And with great power comes great fear of losing it.

# A BEGINNER'S GUIDE TO EVIL IN THE TWENTIETH CENTURY

The next morning was December 31, the big day. I turned on the TV as soon as I awoke.

Midnight had already passed uneventfully in Australia. I looked across the street to the *Washington Post* building, pulled out my binoculars (I knew they'd come in handy), and peered into the windows. There didn't seem to be any commotion. Everything—the streets, the hotel, the air—seemed quiet and still, reassuring yet eerie.

To kill time before the concert, Bianca and I went to the Holocaust Memorial Museum. Looking back on the man-made atrocities that had occurred just fifty-five years ago made the Y2K bug seem benign. All the predictions of the extremists paled in comparison to the concentration camps, mobile death squads, and bloody reprisals of the Nazis and their vision of a new world order.

One of the most unsettling things about Adolf Hitler is that he wasn't necessarily an imperialist, like Napoleon or William McKinley. He wasn't just trying to subjugate other countries. His goal was to cleanse them, to wipe out the so-called weak races and speed the evolution of the human species through the propagation of the Aryan race. And for seven years, he got away with it. Few of the most brutal periods in medieval history—from the sack of Rome to the early Inquisition—were as coldly barbaric as what happened in our supposedly enlightened modern Western civilization.

And though I left the museum with the reassuring message that the world stood up and said "never again" to genocide, it only took a minute of reflection to realize that it happened again—immediately. In the USSR, Stalin continued to deport, starve, and send to work camps millions of minorities. As the bloody years rolled on, genocides occurred in Bangladesh in 1971, Cambodia in 1975, Rwanda in 1994, and in Bosnia in the mid-1990s.

All these genocides occurred in ordinary worlds where ordinary people went about ordinary business. The Jews were integrated into every aspect of the German social and professional strata before the Holocaust. The entire educated class in Cambodia—teachers, doctors, lawyers, anyone who simply wore glasses—was sent to death camps. And as Philip Gourevitch wrote in his book on the Rwanda massacre, "Neighbors hacked neighbors to death in their homes, and colleagues hacked colleagues to death in their workplaces. Doctors killed their patients, and schoolteachers killed their pupils."

This sudden snapping of the social contract had always fascinated and terrified me—not just genocides, but also much smaller-scale instances of group violence, such as the riots in the United States triggered by Martin Luther King's assassination, the 1968 Chicago Democratic Convention, and the beating of Rodney King in Los Angeles. At its root, most people's fear of the millennium had less to do with the loss of electricity than with the snap that could follow if the system broke down.

So what I ultimately learned at the Holocaust Museum was

not "never again," but "again and again and again." Maybe even tonight.

After leaving the museum, Bianca and I met Trisha Yearwood in the hotel lobby to take special buses, which had been swept for bombs by the Secret Service, to the Lincoln Memorial. Police escorts raced us through the streets, lights flashing, sirens Dopplering. When we arrived at the concert, we went through a battery of metal detectors, security questions, and bag searches. I wondered if the extremists had won, making us—the so-called normal people—just as paranoid as they were.

"Under Clinton's seat," McCain's assistant told me, "there's a trapdoor and a staircase, which leads to a limousine that always has its engine on, in case there's trouble."

McCain was assigned a seat directly behind Clinton, but rather than seeing this as a choice seat in the emergency exit aisle, his assistant feared it would be a Republican catastrophe. "I have to get him moved," he said. "If their photo is taken together, it could reflect badly on Senator McCain's campaign."

This was why only the uptight, small-minded kids in school got involved in politics, I thought. It's not about changing the world. It's still about what lunch table you sit at.

We shuffled to the greenroom to join the performers, politicians, and undercover agents. China had made it past midnight without a glitch—and in a celebration so beautiful it seemed like the ruling culture of this next century would once more be the East. Other than the millennium clock shutting down on the Eiffel Tower, Europe was clear as well.

Across the world, political situations seemed to be resolving. One hundred and fifty-five hostages on an Indian Air jet hijacked to Afghanistan were freed. Boris Yeltsin stepped down as Russia's president. The NASDAQ hit a record high. More good omens for a new millennium. I could feel a collective sigh of relief around me. It looked like we were going to be okay. Not just today, but forevermore.

Compared to China's costumed and colorful blowout, America's millennium concert seemed bland and unfocused. At 11:45, as I stood in front of the stage trying to think of something special to do for the great anticlimax, Bianca ran over with a plastic cup of champagne.

"Trish wants us to join her backstage," she said. "She's s'posed to sing 'America the Beautiful' at 12:05."

We rounded the scaffolding, showed our passes, and walked through the rigging to arrive stage right. Trisha, flanked by Quincy Jones and Kris Kristofferson, stretched out a boyfriendless hand and grimly accepted a cup from Bianca. Several men in black suits stood directly in front of us. I looked at their backs and noticed that just ahead of them—less than twenty feet away—President Clinton stood, waffling into the microphone.

He was saying something about the unique responsibility we had to lead the world. "The sun will always rise in America as long as each new generation lights the fire of freedom," he concluded. It sounded good. But, really, when you think about it, what the fuck does that mean?

Before any of us had time to think about it, the countdown began.

5-4-3-2-1. *Snap.*

I took a photograph in Clinton's general direction that looked like this. The Secret Service men didn't even turn their heads:

The lights didn't go out. The power didn't shut down. The world didn't end. Overhead, the artificial sun flickered and sputtered to life, much to the relief of many stagehands. We were all still here, shouting and hugging, engaged in the same annual ritual of forced festivity.

Atop the Lincoln Memorial, I saw a line of Secret Service men with night-vision goggles, aiming rifles into the crowd. And I remembered my pledge to avoid spending New Year someplace with guns. I guess I'd broken my promise.

I wondered what Bob Rutz and his band of psalm-quoting survivalists were thinking at their last skate. Had they wasted their time and money building their Arkansas compound?

I was glad I hadn't fallen for the survivalists' panic. For a moment, they'd almost had me scared. But at least I was in good company. Judging by the conversations I'd had here, they'd almost had everyone scared.

When the concert ended, we boarded a bus to the White House. At various checkpoints, we were sniffed by dogs; our IDs and social security numbers were examined; and we were frisked by a blond, mustachioed man, who was hopefully part of the White House security team and not just some random pervert.

The first person we saw when we entered the building was Muhammad Ali, perhaps a perfect symbol for the decade to come—a former powerhouse battling a degenerative disease. Next to him Martin Scorsese, Jack Nicholson, and Robert De Niro were talking in a huddle. To me, these were famous people, embodiments of the American dream. But to Bianca, emboldened by alcohol, they were bowling pins. She threw me against the wall, breath reeking of champagne, and drawled, "You're my guest. Don'cha run away from me."

So I walked away.

It was my first time in the White House, and the middle two floors of the building were almost completely open for guests to wander through. The first illegal act I committed was stealing this hand towel from the bathroom, solely because it had a presidential seal on it:

In truth, the White House never held much mystique for me. The kids who were interested in politics, who formed the Young Republicans Club, who volunteered to work in local elections, were the boring ones. Powerless at the playground and parties, they were only able to feel self-important in a controlled environment like politics with specific rules and prestigious titles.

I know because I was one of those uncool kids. I was in student council. I volunteered to staff polling places on election days. I even joined the Young Republicans Club. I didn't really know the difference between Democrats and Republicans. I just wanted to belong to something and to bond with the slightly cooler uncool kids I so self-respectlessly followed.

As I listened to Barney Frank, the only openly gay congressman, urge a nearby gay couple to get on the dance floor to send a message to his fellow politicians, I felt a large claw grab my shoulder. "We're going home," Bianca hissed in my ear, upset that I'd slipped away from her.

An hour later, we were back on the private plane, leaving as suddenly as we'd arrived. As everyone else tried to sleep, Bianca—more drunk and aggressive than I'd ever seen her—kept turning my swivel chair to face hers and talking dirty. "I want you to, y'know, cum in my face. I've never done that before. What does it taste like?"

I told her it was kind of like eating lychee nuts. I'm not sure why that image came to mind.

As the New Year's sun rose behind the plane, Bianca thrust across the seat, lowered her dress, and stuck her breasts in my face. This, I thought, was an uncomfortable way to begin a new millennium. I just wanted to escape.

When I finally made it home, I checked my e-mail before going to sleep. There was a message from one of my new doomsdayer friends: "Hi, Neil Strauss, it's Tim Chase. This new president Putin in Russia? Interesting that he became the Russian president as the millennium changed. He could be the one, the Prince of Darkness."

And I knew then that I'd made the right choice. There will only be one millennium party at the White House. But there will always be another apocalypse.

# THE MAGIC OF LIFE: NOW YOU SEE IT, NOW YOU DON'T

When I was twelve, I used to wonder if I would live to see the year 2000. It seemed so far away. I'd sit in class and count on my fingers, trying to calculate how old I would be. In my head, I'd beseech God, "Please don't kill me before the year 2000. I want to see what it's like."

And he spared me.

When I was fifteen, I used to wonder when I'd lose my virginity. I used to lie in bed, thinking about what it would be like to finally feel the naked flesh of a girl. And I'd beseech God, "Please don't kill me before I experience sex. I want to see what it's like."

And he spared me.

When I was nineteen, I used to wonder what Europe was like. I'd sit in the college library, daydreaming about the cathedral of Notre Dame, the canals of Venice, the Eurorail, and Swedish women. And I'd beseech God, "Please don't kill me before I go to Europe. I want to see what it's like."

And he spared me.

Today, I wonder what it will be like to become a father. Every night, when the nagging drone of unfinished work dies in my temples, I hope and pray that I don't die in some sort of accident before I get to bring life into this world and watch that small, helpless being grow into adulthood.

Yet on every highway, there's a drunk driver hurtling at 80 miles an hour in two tons of steel. In every neighborhood, there's a thief armed with a deadly weapon. In every city, there's a terrorist with a bloody agenda. In every nuclear country, there's a government employee sitting in front of a button. In every cell in our body, there's the potential to mutate into cancer. They are all trying to kill us. And they don't even know us. They don't care that if they succeed, we will never know what tomorrow holds for us.

The tragedy of life—robbing it of its fullness and brilliance—is the knowledge that we might die at any moment. And though we schedule our lives so precisely, with calendars and day planners and mobile phones and personal information management software, that moment is completely beyond our control.

Death is a guillotine blade hanging over our heads, reminding us every second of every day that this life we treasure so much is no more important to the universe than those of the two hundred thousand insects each of us kills with the front of our car every year.

Nature knows no tragedies or catastrophes. It knows no good or evil. It knows only creation and destruction. And one can never truly be happy and free, in the way we were as children before learning of our mortality, without at some point confronting our destruction. And all we can ask for, all we can hope for, all we can beseech God for, is to win a few battles in a war we will ultimately lose.

# STEP 2: **SEPTEMBER 22, 2000**

LESSON 6

# WHY THEY HATE US BUT LIKE OUR MOVIES

The postcard depicted a bearded soldier with his pants around his ankles. Bent over like a football center in front of him was Mickey Mouse. It was clear from the pained expression on Mickey's face that he was being forcibly entered.

Above the pair in red, white, and blue lettering were the words FUCK THE USA.

It wasn't the kind of image I expected to see on a postcard rack in the tourist center of Belgrade, Serbia. So I bought it:

I moved on to another vendor selling T-shirts, ceramics, and tourist paraphernalia. On his rack was a similarly illustrated postcard of a Serbian soldier. This one was urinating on an American flag. The caption: U CAN'T BEAT THE FEELING!

So I bought that too, along with these ten other postcards attacking America and NATO:

In the month before Y2K, I'd learned about threats to America from within. Now I was learning about threats from the outside. But I didn't take them seriously, any more than I did the survivalists. In my naïveté, I was actually excited to add these items to my growing anti-American propaganda collection.

I'd started the collection nine months into the new millennium while traveling in Iran with my brother and father, who was studying the Silk Road, the ancient trading route connecting Asia to Europe and North Africa.

Driving into Tehran from the airport, we passed a building with an American flag painted on the side. The stars had been replaced by skulls and the stripes were trails from falling bombs. Emblazoned in large letters over the flag were the words DOWN WITH THE U.S.A.

It was the first time I'd been in the presence of such a striking display of hatred toward my own country. But instead of feeling fear or shock or outrage, as I would have expected, I was struck by an emotion I couldn't pinpoint. Because it had been painted many years ago and then abandoned to the elements, the mural seemed like a page from an outdated history book I wasn't supposed to see:

As we entered Tehran, we passed a brick wall topped by a black fence. Along the wall was more faded anti-American propaganda, including graffiti of the Statue of Liberty, her green feminine face replaced by an evil, grinning skull. Struck by the brutality of the image, I took this photo from the car:

Behind the fence was a building I never thought I'd actually see: the U.S. embassy. Looking at the disappearing graffiti on the wall and the blossoming green foliage on the other side, I thought about how long it had been since the hostage crisis of 1979, how times had changed, and how we were in a new millennium of peace, prosperity, and democracy. Under the leadership of the reformist president and former minister of culture Mohammad Khatami, Iran no longer wanted to destroy America. It wanted to open up and modernize.

In the Golestan Shopping Center, women wrapped in burkas shopped for designer jewelry. Though the only skin showing was the front of their faces peering out from beneath black chadors, at least one in twenty of those faces had a bandaged nose from recent plastic surgery. My cab driver later told me that Iran was the world capital of nose jobs, proving that even in a culture like this, a woman's vanity could not be kept down.

It seemed as if Iran was slowly falling in step with the West. Although there was no Kentucky Fried Chicken in Tehran, there was a Kabooky Fried Chicken that looked suspiciously the same. Although there was no McDonald's, there were dozens of fast-food restaurants selling burgers called Big Macs.

When we visited tourist attractions, markets, and mosques, men stood up and greeted us. "We're so glad Americans are visiting again," they would say. Then they'd often ask, "You don't hate us anymore?" After we reassured them, they'd continue hopefully, "Do other Americans feel the same?"

It was as though a family rift had been resolved and resentment over an incident two decades ago was finally fading. It was further evidence that the survivalists hiding from the world in their hillside retreats were wrong. We were entering a new era of tolerance and understanding. There was nothing to be afraid of.

As we traveled through Tehran, Shiraz, Isfahan, and the ancient city of Persepolis, the only time we heard anything negative about Americans was when we gave a painter the thumbs-up sign and learned that the gesture actually signifies *fuck you* in Iran.

Perhaps people everywhere are the same; only the symbols change. The problems occur when people believe their symbols are the right ones and everyone else's are wrong.

Maybe that's why the Serbians had so many anti-American postcards. Though the Clinton administration's decision to bomb noncivilian targets had helped to end the genocidal Slobodan Milosevic regime, the U.S. had taken the moral high ground. The problem with that is it leaves someone else on the moral low ground, and being put down there is so repellent to human nature that the only solution is to claim a different moral high ground yourself. This is how hatred is created: two different groups, each insisting they're on the moral high ground.

One day, as we walked to the Imam Mosque in Tehran, we noticed tanks rolling down the street. They were followed by missile launchers, antiaircraft guns, and squads of soldiers. This was more like the Iran I'd imagined as a child.

"What's going on?" I asked after finding a soldier who spoke English.

"It's Sacred Defense Week," he told me.

"Is there a war?"

He explained that it was the anniversary of the eight-year-long Iran-Iraq War, and the military was showing the people and the Ayatollah it was still capable of defending the country. "When your biggest enemy lives next to you," he continued, "it's necessary sometimes to make your people feel safe."

I felt so ignorant. I may have known everything about the entertainment world, but I knew nothing about the real world. I had no idea that Iran's biggest enemy wasn't the United States but Iraq.

I felt lucky, in that moment, to be alive in an era when I could safely travel just about anywhere in the world as an American without encountering enemies. The most forbidden place was Cuba, but even the embargo there seemed like a vestige of a dead Cold War.

As Francis Fukuyama had foretold in his essay, the world seemed to have advanced beyond wars over religion and nationalism. These conflicts were limited to more primitive societies, which just needed time to catch up. Norman Angell had written in a similar work, *The Great Illusion,* that war was becoming obsolete in the face of a modern world full of multicultural, polyreligious societies that were economically dependent on each other.

Of course, Angell's book about the end of war in the modern world was written in 1911. Three years later, World War I broke out in Europe. So perhaps all that Angell and Fukuyama and, it would turn out, myself were feeling was the quiet before the storm, buoyed by a resolute human optimism and enough vanity to believe we were living at the end of history—in much the same way that World War I was known as the War to End All Wars, though in truth it was just the war that made the next war possible.

And so, after I returned from Iran with my graffiti photos and official government stamps like these—

—I officially began collecting anti-American propaganda. Much of it came from Russia during the Cold War, China during the Cultural Revolution, and North Korea today, like this poster:

But what was most exciting was finding unexpected sources of vitriol, like this late-nineties advertisement by an Indian jeans company, with the slogan at the bottom NOT MADE IN AMERICA. THANKFULLY:

America is the police
of the world.
Don't wear its uniform.

SUNNEX

NOT MADE
IN AMERICA. THANKFULLY.

Available in Parallel, Relax and Slim fits. Manufactured by: Annapurna Apparels Pvt Ltd, Mumbai 400 086. Tel: 514 9603, 515 2177. Fax: 516 5899.

I collected them for the same reason I wanted to spend New Year's Eve with Bible-thumping prophets of doom—because I didn't take them seriously.

After all, we were America, where all the lessons of the past had supposedly coalesced into perfection. We had the best movies, the best music, the best government, the best opportunities, the best lives. We didn't have to invade countries. Instead, we

opened a McDonald's in their town square and played *Die Hard* in their theaters and put the Backstreet Boys in their stadiums. And the more they ate our food, the more they admired our action heroes, the more they hummed our songs, the stronger we became.

Of course that created resentment, which expressed itself in the form of the propaganda I collected in much the same way a singer confident in his talent makes a collage of bad reviews by hack writers and hangs it on his wall with pride.

I was too ignorant at the time to realize that it wasn't our burgers but our policies that were responsible for this resentment—and that its consequences would be fatal.

Unbeknownst to me, another book was released as I was traveling through Iran in my rose-colored glasses, and it wasn't about the end of history. It was about the end of an empire. Written by Chalmers Johnson, a former consultant for the CIA, the book was *Blowback,* named after intelligence jargon for the unintended repercussions of covert foreign operations.

"The evidence is building up that in the decade following the end of the Cold War," Johnson wrote, "the United States largely abandoned a reliance on diplomacy, economic aid, international law, and multilateral institutions in carrying out its foreign policies and resorted much of the time to bluster, military force, and financial manipulation."

Johnson went on to warn that "the by-products of this project are likely to build up reservoirs of resentment against all Americans—tourists, students, and businessmen, as well as members of the armed forces—that can have lethal results."

If I knew then what I know now, I would have realized the obvious—that my collection was a symptom not of open-mindedness, but of the exact American naïveté and arrogance that leads others to hate us in the first place.

Because when I look at these stamps, T-shirts, and posters today, they don't seem so funny anymore:

**STEP 3:** ********* **, ****

## LESSON 7

# THE WORLD'S OLDEST SURVIVAL MANUAL

After calling friends in New York to make sure they were okay—only to be confronted with an ominous busy signal every time—I raced out of the house, drove to the nearest gas station, and filled my tank. With planes down in three states so far, it wasn't clear yet that the attacks were over.

Though I imagined panic in the streets and long lines at gas pumps as people tried to flee to rural areas, the Arco station near my home was strangely quiet. Afterward, I drove to the grocery store to stock up on water. I imagined pandemonium as families filled their carts with supplies, but instead it was eerily deserted.

Perhaps most people were sitting at home, glued to the television, awaiting more information and further instructions. But my Y2K conversations had taught me that there are only two kinds of people in a crisis: the quick and the dead.

Where the future was uncertain on the eve of the millennium, after 9/11 it was much more certain: there were people out there who wanted to kill us just for being American. My propaganda collection suddenly wasn't so hip and ironic. It was a warning sign.

Just like when man first walked on the moon or ran the mile in under four minutes, all of a sudden anything was possible. If they could hijack planes and blow up the World Trade Center, then they could just as easily slip a biological agent into our water supply or release nerve gas into a crowded airport, subway, school, or theater. It now seemed like common sense to take precautions.

Of course, I was years away from becoming a true nutcase. Back then I was just a reactionary, scrambling like everyone else. A survivalist is prepared beforehand.

After stocking up on water and canned food, I stopped at an ATM and withdrew $200 in emergency money. If there was a national crisis and the power went out, I'd need it to buy more gas and supplies.

When I returned home, I opened the copy of the Bible I'd bought while researching the millennial doomsdayers and stashed the money inside. The page happened to be Proverbs 27: "A prudent man foreseeth the evil, and hideth himself; but the simple pass on, and are punished."

I didn't take it as a sign, though—there's an apt prophecy on nearly every page of the Bible. It is, after all, the original survivalist manual, full of righteous men fleeing floods, fires, plagues, genocides, and tyrants.

Though it took the federal government seventeen months to issue its first clear instructions on preparing for another terrorist attack (sending a nation of would-be survivalists shopping for duct tape and plastic sheets), it took only seven days for the next one to occur.

On September 18, letters containing anthrax spores began arriving anonymously at the offices of government officials and journalists, eventually infecting twenty-two people and killing five. As a reporter at the highest-profile newspaper in the country, I was suddenly that much closer to being a target. So when a suspicious-looking envelope with a handwritten label arrived for me at the Los Angeles bureau of the *New York Times,* I left it unopened.

In that moment, I realized I was no longer a detached observer, chronicling and mocking the paranoid. I was now officially one of them. And so my stack of unopened hand-addressed letters and packages grew from a single envelope to a small pile to a veritable mountain.

We make fun of those we're most scared of becoming.

Fortunately, in case I caught a respiratory infection while traveling in Iran, my doctor had given me a prescription for Cipro, which happens to be the same pill to take in case of anthrax exposure. So I felt vaguely protected. But newscasters also warned that a chemical attack could be next, and I had no protection from that.

So I decided to purchase a gas mask.

Like many others since the attacks, I was haunted by a demon I'd always known about but had never met face-to-face before. It's the same demon that haunts mothers who are overly protective of their children and people who take aspirin before they actually have a headache.

The demon is known by the name of Just in Case. It has many heads. And the more fear you have, the more heads you see.

## LESSON 8

# THE PROBLEM WITH GAS MASKS

I was too late. The shelves of Major Surplus and Survival had already been picked clean. Just minutes before I arrived at the army supply store, a husband and wife had bought the last six gas masks for their family.

Fortunately, one of the first lessons I learned as a journalist was to always go to the source. So I decided to call the distributor that supplies the army surplus stores.

Again, I was too late. "We sold in excess of twenty thousand gas masks in three days," Pamela Pembroke of CORP Distribution's sales department told me. "Even our supplier overseas is out. There's just no more supply." She paused, then added, "If it's going to be that bad, I don't want to be around."

After a few more hours of research, I found an emergency-planning company called Nitro-Pak. This time, I called and asked to speak to the president, Harry Weyandt. It was easy to find his name because there was a thumbnail picture of him on the Nitro-Pak website. He looked incredibly normal, like the proud parent of a high school quarterback.

Within thirty minutes after news of the attacks spread, Weyandt told me good-naturedly, the phones at his company were ringing off the hook. Sales doubled on Tuesday and Wednesday, then tripled on Thursday. "By Friday they went up by seven hundred percent," he continued, "and in the last few days, they've gone up three thousand percent."

"It looks like you're in the right business," I said in an attempt to befriend him before begging for a spare gas mask.

"It's been a wake-up call for all of us," Weyandt said. "We all pooh-poohed Y2K because nothing happened. Anyone who got prepared was seen as foolish. But now people who were prepared are seen as prophets."

I suppose I had to agree with him, especially since buying beef jerky and a Bible made me sort of a prophet by his definition. St. Slim Jim.

"So," I finally asked, "do you have any gas masks left in stock?"

"I called eight of our suppliers, and every single one of them is sold out."

"I figured. It was worth a shot."

"But," he continued, "I was able to eventually find some Israeli civilian gas masks that are being shipped to us now. We're probably the last ones who have any."

"Are those reliable or just collector's pieces?" It seemed too good to be true.

"They're the masks the Israeli government issues to every civilian. They're very popular and easy to use, and they'll give you six to ten hours of clean, filtered air in case of a nuclear, chemical, or biological attack."

Sold to St. Slim Jim.

"The other thing that's been very popular is the Evac-U8 smoke hood," he continued. He knew a sucker when he heard one. "It's the size of a soda pop can, and in the event of—heaven forbid—there's a hood that protects you from heat and flames and takes out carbon monoxide. If you ever use it in an emergency, the company gives you a free one as a replacement. It's only sixty-nine dollars, so that's cheap insurance."

As he spoke, I searched for the smoke hood online. It looked ridiculous:

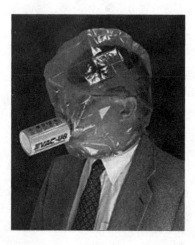

I may have been paranoid, but I wasn't that paranoid. Maybe I'd made a mistake trying to befriend him. I felt like I was talking to a car salesman. The difference is that the car business benefits from optimism and wealth. The survival business benefits from fear and tragedy. It sells not speed, but longevity.

And since life is something I enjoy when I'm not taking it too seriously, I let Weyandt talk me into several additional items that were only slightly less preposterous than the Evac-U8 smoke hood.

A week later, I received two space blankets, a heavy-duty tube tent, two waterproof ponchos, a thirty-six-hour emergency candle, a box of waterproof matches, twenty-four purified drinking water pouches, a first aid kit, an AM/FM radio, an emergency survival whistle, a pair of leather gloves, a random nylon cord, and, just to ensure that every human need was cared for, a small cardboard folding toilet. They were all part of a seventy-two-hour emergency kit, conveniently squeezed into a black duffel bag in case I had to evacuate my home during a crisis.

In addition to the kit, I also ordered a box of twenty-four ready-to-eat meals (MREs, in military parlance), which contained beef stroganoff, chicken stew, cheese tortellini, and other entrées and snacks, all freeze-dried in small packs and made to last roughly seven years unrefrigerated. I had no idea at the time that exactly seven years later, I'd find myself eating those meals.

I brought the supplies to the garage, hoping I'd never have to use them. My precious gas mask, in particular, looked daunting. It was packaged in a cardboard box covered with red Hebrew letters and handwritten numbers. Inside there was a rubber gas mask, a filter secured by a silver sticker tab with more Hebrew writing on it, a piece of white plastic that looked like a miniature toilet seat, a metal cap, and a cardboard ring.

I had no clue how to use it, and the Hebrew instructions weren't helping any. So I called David Orth, an assistant fire chief I'd once interviewed, for pointers.

That's when I learned I hadn't actually bought survival. I'd only bought the feeling of safety.

"A gas mask is iffy, so I wouldn't depend on it," Orth told me as soon as I asked for help.

"What do you mean?" My heart sank. "Do you know how hard I worked to get this thing?"

"The filters are really designed for basic irritant gases and won't protect you against a nerve agent," he continued. "It won't really protect against a biological agent either, because those may not be purely an inhalation issue."

"What other kinds of issues are there?" My chances of survival were shrinking by the second.

"Skin contact with anthrax is a more common cause of infection than inhalation. With sarin gases, also, there's a threat of skin contact. As firefighters, we wear positive-pressure self-contained breathing apparatuses, so I wouldn't use a filtering-type mask as any guaranteed protection."

"So I'm basically screwed?"

"It's a false sense of security," he concluded.

For a moment, I wished I'd bought the Evac-U8 smoke hood. But only for a moment. The company later recalled the smoke hood when it was discovered that it didn't completely filter out carbon monoxide.

When I first relocated to Los Angeles in 1999, my boss at the *New York Times*, Jon Pareles, had advised against the move, warning me about earthquakes and riots. Now, after 9/11, Los Angeles actually seemed like a safer place than New York. Not only was the city too spread out for a single target to immobilize it, but, unlike Manhattan, it had no single building or monument that symbolized the nation.

However, when I called Pareles to discuss the possibility of another attack in New York, he seemed nonchalant.

"Aren't you worried living there now?" I asked. "I mean, it's the terrorists' number one target. And it's a small island with only a few bridges and tunnels for escape. It would be easy for terrorists to shut it down or take it out."

As if the answer were obvious, he replied, "Who wants to live in a world without New York?"

I suppose, then, that there are two types of people in the world: the captains, who go down with their ship, and the rest of us, who jump off with our loved ones.

In the immediate aftermath of 9/11, my plan was to stay on the ship. Not only did I halt my fast-growing collection of anti-American propaganda, but I began to feel, for the first time, a sense of patriotism welling inside me.

I first noticed it when I heard friends from other countries calling Americans obese, rude, or uneducated. Instead of agreeing, I found myself arguing with them. Suddenly, negative stereotypes of Americans seemed not only dehumanizing, but also dangerous.

When I spent six months working in Nashville for the *Times,* I found it odd that people there identified themselves as Southerners. After all, growing up in the North, we never thought of ourselves as Northerners. We were simply Americans.

But after 9/11, I understood why Southerners were so proud. As Northerners, we'd never been marginalized in our lifetime, so we'd never had to unite and prove ourselves to anyone. Now, as Americans, we were marginalized, and it was time to prove to the world that it was wrong about us.

Unfortunately, that's not what happened.

# EFFECTS OF SCHOOLTEACHERS ON PSYCHOSEXUAL DEVELOPMENT

I was eleven years old when I first learned about the Holocaust. I may have heard the word before, but I'd never really understood what it was until Mrs. Kaufman put it in context.

Though Mrs. Kaufman was grayhaired, wrinkled, and probably a grandmother, she had the most tremendous breasts any of us sixth graders had ever seen on a teacher. Even her thick cardigans were unable to conceal their enormity.

I can't remember the name of our history textbook, except that it had a red cover, an ominous thick black swastika in the middle, and, on the inside of the back cover, a detail I added: twenty-five numbered illustrations of different sizes and varieties of breasts.

During the second week of class, Mrs. Kaufman drew a timeline on the board. It began in 1933, with Hitler's appointment as German chancellor and the boycotting of Jewish professional services. Below the year 1935, she wrote "The Nuremberg Laws," which stripped Jews of their citizenship. Now she was at 1938. "Jewish passports stamped with J," she wrote. Then, on the line underneath, she wrote "Kristallnacht" in big letters.

"Can anyone tell me what Kristallnacht was?"

As the class clown, it was my duty to come up with a joke response to every question. But instead, I listened transfixed, imagining myself suffering each successive indignity. I don't know why, but when studying literature and history in school, I never identified with the oppressor—only with the victim. Perhaps because that was also my role in the social pecking order of sixth grade: the small, funny kid who was bad at sports, awkward around girls, and fell over easily when pushed, especially when carrying books.

Penny, a transfer student who was blond and smart and perfect, raised her hand. "It was the night the Nazis rioted against the Jews," she said, the kiss-ass.

"That's right." Mrs. Kaufman nodded approvingly. "Hundreds of synagogues were destroyed, thousands of Jewish homes and businesses were ransacked, and Jews were beaten and murdered in the streets."

We'd already learned, a week earlier, the end of the story: ghettos, concentration camps, gas showers, and a new and ugly word I'd been taught: genocide.

"Why did they stay?" I blurted.

"What do you mean?" Mrs. Kaufman asked. It was impossible to look at anything but her breasts when she spoke.

"When things were getting so bad, why would any Jew stay in Germany?"

"A lot of people didn't think things would get any worse," she answered. "And by that point, it was harder for Jews to leave."

As the Holocaust was retaught with further elaboration in each successive European history class, I found myself more and more amazed that people would willingly remain in a country that was stripping them of their rights and homes.

And I told myself the same thing every time: If that ever starts happening here, I'm not going to wait around, thinking things can't get any worse. I'm getting out, before it's too late.

# STEP 4: **NOVEMBER 3, 2004**

# CALCULATE YOUR SAFETY IF X = TERRORISTS, Y = DISTANCE, AND Z = HOW MUCH YOU'VE PISSED THEM OFF

t's incredibly safe. We've never had terrorist threats or hijackings here. We're like a forgotten country because we're so far from everyone. We don't worry about security like in America. And we don't invade countries on false pretexts and make the whole world mad at us."

Jane, a girl from Sydney I'd briefly dated, was trying to convince me to move to Australia, where the coastline is beautiful, the crime rate is low, and the people are easygoing and friendly.

I racked my brain to think of a hole in her argument. "But you're pretty close to Indonesia. And the terrorist group that bombed the Bali nightclub a couple years ago is there. Doesn't that worry you?"

"When the Bali bombing happened, a lot of Australians stopped traveling to Indonesia," she answered, unfazed. "But as a country we don't have any problems with Indonesia. I think the media and your president like creating fear in people, because when people are in fear, they remain docile. They don't question things and are grateful to the government for protecting them. Over here, it's like a little paradise. We enjoy what we have and don't ask for anything more."

"Well, there's an election coming up here." I found her enthusiasm endearing, though I didn't take it seriously. Not yet. "So let's see what happens. Maybe things will get better."

"If you want somewhere even safer," she said as I hung up, "there's always New Zealand."

# POP QUIZ: RAISE YOUR HAND IF YOU DIDN'T VOTE

On November 3, 2004, the New Zealand immigration service received 10,300 hits from the United States on its website—more than four times its daily average—while calls and e-mails from Americans inquiring about immigration skyrocketed from an average of seven a day to three hundred.

One of those inquiries was from me.

For the first four years of the Bush administration, we were blameless. After all, we hadn't technically elected the president. And back then we had no idea that he would lead us into an unnecessary war, bring the budget from a $236 billion surplus to the highest deficit in U.S. history, strip away civil liberties in the name of national security, and disregard international treaties, the United Nations, and the Constitution.

After the 2004 election, however, everything was different. This time, Bush had actually been voted into office. And the message that sent to every other country in the world was that the people of America condoned his actions. Thus, it was no longer Bush who was stupid in the eyes of the rest of the world, it was us.

The cover of Britain's *Daily Mirror* said it all: "How Can 59,054,087 People Be So DUMB?" (It failed to mention the more than 79 million Americans who were even dumber and, though eligible to vote, didn't.)

For historical-precedent-obsessed Americans like me, though, this was about more than George W. Bush. Just as the attacks of 9/11, though they were far from traditional warfare, showed Americans that war could happen here, the national security clampdown of the Bush administration—though far from actual authoritarianism—showed Americans that fascism could happen here. After all, to make an extreme comparison, even Mussolini and Hitler came to power legally in democratic governments.

And so I lay on my bed that afternoon, stared at the white plaster ceiling, and thought about what I could do. I'd had my chance during the 2000 election: the thirty-five days between the end of polling and the final announcement of the victory were one of those pivotal moments in American history where a single person could have made a difference. And I had a better opportunity than many. As a reporter for the *Times,* all I had to do was fly to Florida, find the right story, and expose it before a decision was made. Instead, like most other Americans, I watched TV and waited for someone else to do it.

Despite my disillusionment, I didn't believe that Bush, Cheney, and Rumsfeld were incarnations of evil, looking to create martial law or a police state. But they were so zealously pursuing such a narrow set of priorities that they were doing more harm than good.

Whatever store of cultural or political goodwill we had acquired over the years was mostly squandered in Iraq. In 2000, 75 percent of Indonesians—the country with the world's largest Muslim population—viewed Americans favorably. After the invasion of Iraq, that number dropped to 15 percent, with 81 percent of the population saying they feared a U.S. attack. Even

South Koreans, who once considered North Korea their biggest threat, were now more afraid of the United States. And across Europe, South America, and the rest of the world, leaders who openly collaborated with our government lost popularity and elections.

The day the results of the 2004 election were announced was the first time I seriously considered leaving America. I felt alienated from the majority of the country, worried about the damage four more years of the same administration would do, and concerned about a backlash from the rest of the world.

Recently, I'd left the *New York Times,* hoping to move on to bigger and better things. But those things hadn't come. And now, more than ever, I doubted myself. At the newspaper I'd been thought of as the young guy, with my finger on the pulse of popular culture. But the election had proven that my finger wasn't on the pulse. I was just feeling the surface of the skin and imagining a heartbeat that wasn't actually there.

In 1990, before the Internet, e-mail, cell phones, and laptop computers were mainstream, Jacques Attali, an adviser to the French president, wrote a book called *Millennium,* in which he predicted that human beings would evolve into technological nomads. Because technology was making work and communication possible in any location, he elaborated, we'd no longer need to stay in one place.

Perhaps my disillusionment was also an opportunity to pick up my laptop and cell phone, leave the rat race, and become a technological nomad. So that night, with my conversation with Jane in Australia echoing in my head, I checked the immigration websites for Australia and New Zealand.

They required foreigners to live there two to three years before granting them citizenship. That didn't sound unbearable. They also offered citizenships to people in "exceptional circumstances" who will provide "some advantage or benefit" to the country. Unfortunately, my credentials—writing books with drug-addicted rock and porn stars—seemed more likely to hurt than help.

Then again, the governor of my home state was a former bodybuilder who'd admitted to using anabolic steroids, attending orgies, and smoking marijuana (in his words, "that is not a drug—it's a leaf"). So maybe there was a chance.

But when all the buildings around you are still standing; when you can flip on the TV at any hour and watch a reality show; when you can go out at night and drink and dance and flirt and eat a cheeseburger in a diner as the sun rises, it's hard to imagine that anything has really changed or ever will.

The price of my hesitation would be high. By the time I was ready to take action, New Zealand had changed its citizenship requirement from three years of residency to five and Australia had increased its minimum from two years to four. I should have paid better attention to the lesson I'd learned from Mrs. Kaufman: the more people want to leave, the harder it becomes to get out.

I realized then why the Jews in Nazi Germany had stayed: They had hope, which can sustain us in the worst of times but can also be the cruelest of human emotions in uncertain times. And I clung to the hope that we were America and if anything happened, our government would protect us.

# ⸻W T ⸻ B⸻ PRACTICALLY INVISIBLE

THE PROBLEM WITH CAMOUFLAGE CLOTHING AND FACE PAINT IS THAT NO MATTER HOW GOOD THEY ARE, YOU ARE STILL SHAPED LIKE A PERSON AND, THUS, EASILY RECOGNIZABLE. THE *GHILLIE SUIT* IS FAR MORE EFFECTIVE BECAUSE IT BREAKS UP YOUR HUMAN OUTLINE, MAKING YOU ALMOST IMPOSSIBLE FOR THE UNTRAINED EYE TO RECOGNIZE.

1. GET A VOLLEYBALL NET, AND CUT OFF ANY BORDERS AROUND THE NETTING.

2. SPREAD A LOOSE-FITTING CAMOUFLAGE ARMY JACKET AND PANTS (OR FLIGHT SUIT) FRONTSIDE-DOWN ON THE FLOOR, AND LAY ONE SECTION OF THE NETTING OVER THE JACKET AND ANOTHER OVER THE PANTS.

3. WRAP THE EXCESS NETTING AROUND THE ARMS AND TORSO OF THE JACKET, AND THE LEGS OF THE PANTS. IF YOU WANT TO BE CAMOUFLAGED FROM THE FRONT AS WELL, REPEAT THIS PROCESS ON THE OTHER SIDE OF THE CLOTHING. ALSO COVER A BOONIE HAT WITH NETTING.

4. SEW EACH KNOT IN THE NETTING SNUGLY TO THE MATERIAL USING REGULAR NON-WAXED DENTAL FLOSS AS THREAD. REINFORCE EACH KNOT WITH A DROP OF SHOE GOO.

5. GET EIGHT YARDS OF BURLAP IN A COLOR THAT MATCHES THE ENVIRONMENT YOU'RE TRYING TO BLEND INTO (OR 600' OF JUTE TWINE), AND CUT IT INTO QUARTERS. LEAVE 1/4 ITS NATURAL COLOR, THEN DYE THE OTHER 3/4 USING OTHER, SIMILAR SHADES FOUND IN THE ENVIRONMENT (SUCH AS OLIVE DRAB, LIGHT BROWN, OR LIGHT GREEN).

6. CUT THE BURLAP INTO STRIPS ROUGHLY ONE INCH WIDE AND TWO FEET LONG.

7. TIE MULTIPLE STRIPS TO EACH SIDE OF EACH SQUARE OF NETTING, MAKING SURE EACH STRIP OVERLAPS. FIRST ADD THE STRIPS THAT ARE THE COLOR THAT BEST MATCHES THE ENVIRONMENT, THEN ADD THE REST OF THE STRANDS.

8. MAKE SURE THE COLORS ALTERNATE IN A RANDOM PATTERN, AND LEAVE SPACE OVER YOUR EYES SO YOU CAN SEE. YOU MAY WANT TO SEW LEFTOVER STRIPS INTO A SCARF TO BREAK UP THE OUTLINE OF YOUR NECK. WHEN YOU'RE FINISHED, YOU SHOULD LOOK LIKE A CROSS BETWEEN THE HEAD OF A MOP AND COUSIN ITT.

9. FOR ADDITIONAL CAMOUFLAGE, YOU MAY WANT TO TOUCH UP THE SUIT WITH SPRAY PAINT, USE A METAL-TOOTH COMB TO FRAY THE BURLAP, ATTACH FOLIAGE FROM THE AREA IN WHICH YOU'LL BE HIDING, AND ROLL IN THE DIRT THERE.

TO BE CONTINUED...

# STEP 5: AUGUST 29, 2005

# LESSON 12
# MENTAL HEALTH GUIDELINES FOR AIR TRAVELERS

*I*'m a runner.

The security line stretched across the second-floor balcony of the terminal, wound around the check-in counters downstairs, then continued for another hundred feet outside the door. It was clear that it would be at least an hour and a half before any of us arrived at our gates.

I was on my way to New York to promote a book I'd written, *The Game*. Though I was worried no one in the media would care about my adventures with a cabal of pickup artists, fortunately my schedule was packed with press: *Good Morning America*, *The View*, *Anderson Cooper 360⁰*, and dozens of radio shows across the country. Ultimately, most of those shows would be canceled or pre-empted. A new catastrophe was about to shock the nation.

*I ran to my grandmother's house when Michael Zucker threatened to beat me up after school. I ran to a hotel to hide from the wrath of a jealous girlfriend when she caught me talking to an ex. I ran over fallen protesters to avoid the spray of rubber bullets when a police riot broke out during a concert I was covering outside the Democratic National Convention.*

85

As I waited in the security lane, I realized that the America we grew up in is not the same America that exists now. Most tourists have horror stories of customs agents unnecessarily detaining, mistreating, humiliating, or refusing entry to them or their friends. Several travelers, with all their papers in order, have even died in custody.

We've become our own worst ambassadors.

*When danger occurs, the fight or flight instinct kicks in. And since I'm a small person with little fists and a quiet voice, fleeing offers my best option for survival.*

As I neared the checkpoint, I overheard a young, doughy security guard talking about a high-speed lane.

"I don't mean to interrupt, but is there a faster line?" I asked hopefully.

"They're making one," he said.

"Great."

"All you have to do to use it is register for it."

"How do I do that?"

"You go down to a special office, get a background check, get fingerprinted, and get an iris scan."

*I'm not proud to be a runner. A real man, according to action movies and most women, stays and fights. A wimp runs.*

My heart froze. The words "iris scan" filled my mind with images of the movie *Gattaca*, along with dozens of other Orwellian dystopias. The thought of moving to another country came roaring back to me.

*But I'd rather be a living wimp than a dead hero.*

How many baby steps into the abyss would it take before I finally had the courage to climb out?

I imagined the Mrs. Kaufmans of the future writing a timeline of events leading to America's abandonment of the promise stated in its pledge of allegiance, "liberty and justice for all":

**2001**: 9/11 terrorist attacks; 1,200 people arrested and held indefinitely without charge; Bush signs USA PATRIOT Act, allowing the government to secretly wiretap and search the personal records of citizens without a warrant; war in Afghanistan begins; no-fly list created, eventually growing to over a million names.

*As a child, I used to collect War Cards. I'd seen them advertised on TV and talked my parents into ordering them. Every month, a new set of index cards arrived in the mail, detailing different aspects of World War II.*

**2002**: Male immigrants and visitors from over twenty-five countries required to register with the U.S. government; more than thirteen thousand registrants face deportation; Department of Justice allows FBI to spy on religious and political groups without probable cause; Bush doctrine of preemptive war announced; Homeland Security Act passed; Department of Justice memo authorizes torture up to "serious physical injury" in overseas interrogations.

*My parents probably thought the cards were educational. They didn't realize that each War Card formed a new scar in my imagination:*

**2003:** Iraq War begins; Department of Homeland Security established; Operation Liberty Shield detains visitors seeking asylum from thirty-four Muslim countries; Bush continues to centralize and expand power through the unprecedented use of executive privilege and signing statements, which enable him to ignore or reinterpret bills that have passed Congress.

*As I read War Cards about the Allied bombing that destroyed the city center of Dresden, the Battle of Okinawa that left nearly one-third of the civilian population dead, and the Nazi siege of Leningrad that took the lives of 1.5 million residents, I prayed I would never get crushed by the same sledgehammer of history.*

**2004:** Department of Homeland Security begins affixing electronic monitoring ankle bracelets to thousands of illegal immigrants; government outsources domestic intelligence collection to private companies to circumvent laws restricting spying on citizens; US-VISIT system requires all foreign visitors to be digitally photographed, fingerprinted, and checked against a computer database on entry; photos of prisoners tortured in Abu Ghraib prison surface; subsequent Red Cross investigations find evidence of prisoners being sexually abused, set on fire, and forced to eat a baseball at Guantánamo Bay.

Like friendly fire in combat, the government's war on terrorism had wounded its own country instead. And, consequently, every terrorist had won. Even the bungling shoe-bomber Richard Reid had affected the lives of millions of Americans, making it necessary to remove our shoes every time we pass through airport security.

When I turned eighteen, I received another type of war card in the mail. The government sent me a white postcard with a picture of a birthday cake on the front. On the back I was ordered to report to my local post office to register in case a draft was instituted. I went to sleep countless nights over the next seven years hoping our country wouldn't get swept into another major war. I didn't want to end up as a statistic on a War Card of the future.

At the security conveyer belt on the way to New York, I did the airport shuffle—removing my shoes, belt, watch, sweater, and computer—and placed them in trays. That was when I noticed the posted sign: PLEASE BE AWARE THAT ANY INAPPROPRIATE JOKES TO SECURITY MAY RESULT IN YOUR ARREST.

This seemed like more than just a violation of the First Amendment. It was an assault on my sense of humor. Warning people that all jokes would be taken literally would have been just as effective as actually making them criminal.

On a previous flight, I recalled seeing a middle-aged Hispanic man in handcuffs led away roughly by three officers. When I asked a stewardess what had happened, she explained, "He made an inappropriate comment to TSA officials. They don't have much tolerance for things like that anymore."

I felt my identity as an American—based on the lack of this type of state control over individuals—slipping out of my grasp.

*Because of the War Cards and the draft, the government became like John Wayne Gacy in my mind. It was yet another force that could rip my life from me without giving me any say in the matter—and with no regard for who I was or whether I was a good or bad person.*

The only thing that reassured me as I sat disillusioned on the five-hour flight was the knowledge that this wasn't the first time in history when Americans had lost freedoms in times of conflict, with dubious results.

In 1798, on the verge of war with France, John Adams passed the Alien and Sedition Acts, making it illegal to publish criticism of the government and giving authorities nearly free rein to deport foreign residents. During the Civil War, Abraham Lincoln suspended the right of habeas corpus and imprisoned over 13,000 suspected traitors without a trial. In response to a series of anarchist bombings during and after World War I, Woodrow Wilson ordered the arrest of 10,000 alleged radicals, deporting any who weren't citizens. Under Franklin D. Roosevelt, 120,000 Japanese–Americans were sent to internment camps during World War II. And during the Eisenhower administration, more than 10,000 alleged Communists were blacklisted, imprisoned, or fired from their jobs.

*If I were ever drafted or hunted by the government, I told myself, I would run. I used to wonder whether Vietnam draft dodgers were happy in Canada or if they missed home. When villains in movies raced for the Mexican border, I always hoped they'd make it to freedom. And I was amazed that Roman Polanski was able to avoid the legal repercussions of having sex with a minor just by escaping to France.*

So perhaps the natural inclination to want both freedom and security from our government is too much to ask. Especially considering that, according to a Rasmussen Reports poll, these presidents (with the exception of Wilson) are among the ten most popular in U.S. history.

*Foreign countries began to represent safety to me. If things ever got bad in America, I knew there was always somewhere else to go where I could have new experiences and meet new people. War*

*was for people who cared about power. Peace was for those who cared about life.*

But what happens when your government gives you neither freedom nor security?

I would discover the answer to that question, as would every other American, when I landed.

# HOMEWORK: FIND THE LAST STRAW

Whatever doubt remained about getting a backup citizenship was extinguished by the news I heard during the ride to my hotel. Hurricane Katrina had just laid waste to New Orleans. Instead of a press tour, I spent most of the next week in my room, listening to reports of bodies floating in the water, elderly people drowned in their homes, civilians shot in the streets, police looting stores, and humanitarian shelters turned into humanitarian crises.

When it was all over, 1,836 people had died.

More than the iris scanning, more than Bush's reelection, more than the Iraq War, more than the destruction of the World Trade Center, this was what shattered every last illusion about my country that remained.

Unlike 9/11, the government had advance notice that a disaster was threatening to hit one of its cities. And still, the greatest country on earth not only failed to take care of its own people but took five days just to respond appropriately. According to the House of Representatives committee that investigated the response, the disaster was too large for the city, the state, the federal government, and the Red Cross to handle. In the end, the report continued, the Federal Emergency Management Agency "did not have a logistics capacity sophisticated enough to fully support the massive number of Gulf coast victims."

And with a population of half a million, New Orleans is a relatively small city. So what would happen if an equivalent disaster struck a city of 3.8 million, like Los Angeles, or 8.2 million, like New York?

Something changed in me, as it did for many people, in the aftermath of Hurricane Katrina. It felt like the day I first beat my father at arm wrestling. In that moment, I realized that he could no longer protect me. I had to take care of myself.

An anarchist is someone who believes that people are responsible enough to maintain order in the absence of government. That week, I realized I was something very different: a Fliesian. I began to subscribe to the view of human nature depicted in the William Golding novel *Lord of the Flies*. After reading reports of the chaos, violence, and suffering in New Orleans, it became clear that when the system is smashed, some of us start smashing each other.

Most survivalists are also Fliesians. That's why they stockpile guns. They're planning to use them not to shoot enemy soldiers, but to shoot the neighbors trying to steal their supplies.

With almost every one of my book interviews canceled, I sat in the hotel room all week, fixated on the news. In other stories, more pictures of American troops torturing prisoners at Abu Ghraib were surfacing; the value of the American dollar continued to plummet; Osama bin Laden still hadn't been caught; angry Palestinians were setting the West Bank town of Taybeh ablaze; civilians in western Sudan were being killed and raped daily in a genocide the world wasn't doing anything to stop; and President Bush was threatening Iran the same way he'd once threatened Iraq.

It felt like a powder keg had been lit.

It might blow tomorrow, perhaps a year from now, maybe in ten years. No one knows when exactly. But it will definitely blow.

On my last night in town, my friends threw a small party for me at Tao, a bar near my hotel, to celebrate the release of the book.

Eventually the cast dwindled to just five of us, sitting around a small table. Among them were two of my closest friends: Zan, a gregarious drinker and ladies' man who'd flown in from Vancouver, and Craig, a heavyset Internet entrepreneur who devoured science magazines with the same excitement other men read *Playboy*.

"Have you seen the new documentary about the Enron collapse?" Craig asked, unbuttoning the jacket of his baggy white suit. A tireless orator, Craig enjoyed turning me on to movies, music, and ideas in the hope that I would write about them. "It more or less proves that the California energy crisis was completely faked to drive Enron stock prices up. They play actual phone recordings of traders calling a California power plant and ordering it to shut down. Those traders didn't care about making millions of people suffer in order to make a little more money. And that's the problem with America. The people don't matter anymore."

Like me, Craig was a runner. Except he had a different way of running. Where I wanted to avoid dying of unnatural causes, he wanted to avoid dying of natural causes.

In a few months, he was going to visit the Alcor Life Extension Foundation, a cryonics lab, where he planned to sign up to be frozen and preserved in the hope that a future generation would be able to thaw him. I had promised to accompany him—as a friend and, he hoped, as a potential freezermate.

"If there were another terrorist attack in America—even deadlier than the World Trade Center—would you be surprised?" I asked Craig and my remaining friends.

Every one of them answered no.

"I'd be surprised if there *wasn't* another attack," added Craig.

"And if it happened, do you think the government would be able to evacuate you and restore order?" I pressed.

"Definitely not," Craig said.

"So if the government couldn't save one small city from a disaster it knew was coming, then how is it going to save all of us when the shit really hits the fan?"

A babble of alcohol-loosened voices clamored with predictions about the next threat, but no one disagreed. Craig listened silently, preparing his next argument.

"Let me ask you all a question then," he interrupted after a few minutes. "Everyone pretty much agrees that terrorists hate America and Americans, and want to do everything they can to undermine or even destroy us, right?"

"That's definitely one of their priorities," I agreed.

"Most people think that they want to do this by terrorizing us—by blowing things up and making us scared to leave the house."

"That's why they're called terrorists."

"But have you ever stopped to think maybe that's not their plan? Osama bin Laden is not as stupid and uneducated as most Americans believe. Maybe his plan is to destroy our economy. Because that's the only way to truly put an end to America as we know it." He put down his beer and paused to let his words sink in. He was in his element now. "And our government played into his plan perfectly, starting wars that cost hundreds of billions of dollars and have no end in sight."

"You should just come to Canada," interrupted Zan. "You can stay at my house. Hotel Zan."

Craig ignored him and continued. "That's why he bombed the

trains in Spain. He wanted to scare our allies into withdrawing from Iraq, so we'd have to shoulder the financial burden of the war alone."

"Come to Canada," Zan repeated. "Nobody hates us. Nobody thinks we're stupid, except maybe Americans. Plus we have free health care, stronger beer, and you can get the good codeine Tylenol without a prescription."

"And he's winning," Craig went on, glaring at Zan as if daring him to interrupt again. "Our economy is dying. Look." He held out a pudgy hand and hit his pointer finger. "We have a seven-hundred-billion-dollar trade deficit." Then he hit his middle finger. "A seven-trillion-dollar debt." Now his ring finger. "A recession." Finally, his pinky. "And inflation. Gas prices went up forty-one cents the other day. People are getting angry."

"I'm out of here," I blurted. I'd hit my breaking point. It was time to make good on the promise I'd made in Mrs. Kaufman's class, the promise I'd made while reading War Cards, the promise I'd made after receiving my draft postcard, the promise I'd made when Bush was reelected. It was time to find a safe haven overseas. "I'm getting the fuck out of here."

"Stay with me," Zan offered.

It was the most depressing book release party I'd ever been to.

If any kind of cataclysm should happen in America today, considering that 88 percent of the population doesn't even have a U.S. passport, most people would be trapped here. But I wouldn't have to panic when other countries closed their borders to fleeing Americans, nor would I end up stranded in a refugee camp at the border like so many others who'd been sucker-punched by history. With a second citizenship, I'd already be preapproved to live in another country.

And if I ever found myself in a situation where terrorists were kidnapping or executing American hostages, I could use my sec-

ond passport to prove I wasn't a U.S. citizen and escape with my life. Even if nothing bad ever happened, I'd be able to more easily do things forbidden to ordinary Americans, like traveling to Cuba.

As soon as I returned to my hotel room, I opened my laptop and searched for lawyers and companies specializing in immigration law, diplomatic passports, and second citizenships. Within a few hours, I'd sent fifteen e-mails and made ten international phone calls, setting off a chain of events that would change my life.

## PART THREE

# ESCAPE

I must now go a long way . . .
I must face a fight that I have not faced before.
And I must go on a road that I do not know.

—*Gilgamesh,* Tablet III, 2100 B.C.

# HOW TO BECOME IMMORTAL IN ONE EASY STEP

On the afternoon of January 12, 1967, James Bedford stopped breathing. As for whether he's dead and gone or not, that's a matter of debate. Because, thirty-nine years later, I was staring at the metal canister in which he was awaiting resurrection.

Among those searching for immortality, Bedford is somewhat of a hero: the first frozen man. A psychologist and author, Bedford tried to cheat death by having himself frozen through an experimental procedure known as cryonic suspension. Some time in the future, after a cure for his kidney cancer was found, he hoped to be thawed and emerge, the oldest man alive. Until then, he will remain at the Alcor Life Extension Foundation in Scottsdale, Arizona, floating in liquid nitrogen.

I was here with Craig, who'd brought me to an open house at the facility to try to convince me to join him in cold purgatory. He didn't want to wake up in the future alone. And though I'd come reluctantly, there was a man here who would give me the key I needed to escape America. A man I didn't particularly care for. A man wearing a t-shirt emblazoned with an image of the World Trade Center in flames.

A slender woman in her midtwenties with long, dyed red hair and large, ratlike front teeth wiped some sort of yellow mucus onto her lab coat as she brought Craig and me to the operating room.

"If the patient has opted for neurosuspension," she explained, "this is where we remove the head."

"Instead of getting a second passport, just get frozen with me," Craig pressed as the woman showed us a travel case containing a pump the Alcor transport team uses to replace a patient's blood with organ preservation solution. "We'll come back when science has solved the problem of death and people have outgrown old-fashioned ideas like nationality and religion."

Craig clearly was not a Fliesian. Fliesians don't believe in better times. They believe that all times are the same—only the names and faces change.

"There are too many unknowns," I told him. "What if the company goes bankrupt and the bank sells everyone at an auction?" Now the rat-faced woman was demonstrating the process of vitrification, in which the body's vital organs are frozen not into ice, which forms crystals that damage the cells, but into a glass-like substance. "What if your brain is stuck the whole time in some horrible nightmare? Or, even worse, if it's actually awake inside one of those canisters?"

Craig had a well-reasoned answer for everything. Yet as much as I'd like to live forever, I don't like the idea of dying during the in-between period. If the Mujahideen forces ever invaded, the Alcor Life Extension Foundation is probably one of the first western abominations they'd burn to the ground.

When the lab tour ended, Craig and I were ushered into a parking lot outside the building, where the Alcor staff had set up a makeshift picnic for prospective customers. Übernerds, aged hippies, and half-mad scientists sat at folding tables, eating cheeseburgers. Most were wearing silver bracelets that instructed anyone who found them dead to "rush Heparin and do CPR

while cooling with ice to 50 degrees Fahrenheit. No autopsy or embalming." They seemed like poor ambassadors to the future.

At a table in the center of the parking lot, a man with a natty beard trailing halfway down his chest was holding court. He talked rapidly, as if by squeezing twice as many words into each second, he'd double his lifespan.

His name was Aubrey de Grey, and he was a hero to these people. In much the same way scientists have found cures for anthrax and syphilis, he had been working to find a cure for a condition that kills some 100,000 people a day: aging. Just the month before, he'd received a $3 million–plus donation from Peter Thiel, the cofounder of PayPal.

"It's about eliminating the relationship between how old you are and how likely you are to die," he was telling a man and a woman who looked like they could use his help. The woman had long brown hair with strands of gray and was wearing a loose-fitting T-shirt advocating hemp use. The man had shaggy brown-gray hair, a beard that rivaled de Grey's, and a shirt with the words INSIDE JOB printed beneath a photo of the World Trade Center in flames.

"Within our lifetime," de Grey continued, "I'm fairly certain it will be possible to live thirty extra years."

Of course, that's if you die of not-too-unnatural causes. If you perish in a fire or a plane crash or a bombing, no amount of life-extension therapy and cryonic preservation will save you.

"You're wasting your time with these people," the man in the Twin Towers shirt said, turning to face Craig and I. He scratched the side of his long, porous face with the automatic movements of a dog irritated by fleas. "They're more worried about what they're going to do after they die than what they're doing with this life."

I looked at his wrist. There was no bracelet. His wife, also braceletless, was taking twist-ties off three baggies in front of her. One contained baby carrots; the second a thick brown bar that seemed barely digestible; and the third a stick of some sort of butter substitute. In her lap was a canvas carrier containing a small, nervous dog.

"There are more important things to worry about," the man continued, scratching again. "I'm wearing this shirt to send a message: 9/11 was absolutely controlled demolition." His wife nodded in agreement as she pulled a baby carrot out of a baggie and began spreading congealed buttery substance on it. De Grey wisely slipped away, leaving us alone at the table with the couple. "One hundred percent. Let me explain why. The press said the second tower pancaked down in eight point five seconds. That would be impossible."

I watched his wife continue coating her carrot obsessively. "Bush's younger brother and his cousin had the contract for the World Trade Center . . . World Trade Center Seven was not hit by a plane . . . Four hours later it collapses on its own." Craig escaped to buy cryonics life insurance. (To pay the cost of his preservation and storage, all he had to do was sign his life policy over to Alcor.) I was alone now. "Osama bin Laden was employed by the CIA."

Though I have my paranoid moments, I don't buy into conspiracy theories as involved, risky, and unsubstantiated as the ones he was outlining. A true Fliesian knows that large groups of people don't keep secrets that well, especially if leaking it can either bring them glory or hurt a competitor.

The person who convinced me of this was President Lyndon Johnson, who recorded hundreds of his phone calls in the White House. When I first heard him in the documentary *The Fog of War* discussing withdrawing from Vietnam, and basing decision

after decision solely on whether it would make him look bad, I officially retired my conspiracy theory card.

As the bearded man continued his inside-job rant, I waited for a pause so I could extricate myself. But this man did not pause. He seemed willing to talk to anyone who wanted to listen—and, evidently, anyone who didn't. As he tried to make a case for income tax being illegal, his wife grabbed a pickle chip and started spreading butter substitute on it. I found her eating habits much more fascinating than the sound of his voice—until, suddenly, I heard the word "offshore."

My ears buzzed. My heart raced. Blood rushed to my head. Now he had my attention. "The entire fabric of the United States is dependent on a reciprocal system," he was saying. "Should it be interrupted, or should it collapse, there would be mass chaos because the average American is incapable of doing anything for himself. Now is the time to move everything you own offshore."

Suddenly, I felt like a man in possession of a metal detector that had just started beeping. I would never have guessed that this conspiracy nut might have the answer to the question I'd been wrestling with for the past year.

Until then, every path I'd tried to take out of the country had led to a dead end. Of the twenty-five immigration lawyers and organizations I'd contacted after Katrina, fifteen of them never returned phone calls and e-mails, and two of them said it would be months before they could offer me a consultation. A Japanese lawyer, Yoshio Shimoda, told me I'd need to live in Japan for five years and give up my U.S. nationality in order to be a citizen there. And Camila Tsu, a Brazilian lawyer, told me her country required four years of residency, fluency in Portuguese, and relinquishing my U.S. citizenship.

Not only were these time commitments too long, but I wasn't

about to give up my U.S. citizenship. Despite its faults, I still love America. My friends are here, my family is here, and so is Manhattan, Hollywood, Chicago, Austin, most of New Orleans, the national parks of Utah, Kauai, the dry-rub ribs of Memphis, the juke joints of the Mississippi Delta, the Pacific Coast Highway, the Carlsbad Caverns, Clint Eastwood, and the Titan Missile Museum in Arizona, where you can actually turn the launch key in a deactivated nuclear bunker.

An attorney in Rome told me that Italy wouldn't require me to give up my U.S. citizenship, but I'd need to reside there ten years to get a passport. And a lawyer at Lang and Associates informed me that Costa Rica—one of the few countries without a military—has a seven-year residency requirement.

My favorite response, however, came from Dan Hirsch of Hirsch & Associates. "We require a $3,000 retainer fee and a signed retainer agreement," he wrote in an e-mail. "You do not really need my help, but if you want to use it, send me the retainer and the check."

The only glimmer of hope came from three companies, which each suggested trying Guyana. One of them, P&L Group, claimed there were hundreds of thousands of American citizens already living there.

So I did a little research: On the Atlantic coast of South America. Semidemocratic republic. Former British colony. English-speaking inhabitants. Former site of the Jonestown massacre. Sounded nice.

The only problem was that Guyanese passports were among the world's least-credible travel documents because there were too many counterfeits and altered ones floating around. So I was right back where I started.

Until now.

"Do you know," I asked the man in the World Trade Center

shirt, my voice shaking, "the quickest way someone like myself could get citizenship in another country?"

As soon as the words left my mouth, I imagined black vans full of dark-suited Homeland Security agents pulling up and arresting me for sedition. I watched carefully for his reaction, hoping he wouldn't ask too many questions.

"You should check out the Sovereign Society," he said slowly, after some thought.

"What's the Sovereign Society?" I asked greedily.

"They teach people how to be independent of their government."

I didn't like the way he spoke. I didn't like the way he looked. I didn't like the way he smelled. But I would be forever in his debt. For he had just given me the clue I'd been looking for.

Perhaps this man who I disliked so much—this conspiracy theorist in the socially unacceptable shirt—and I were not so far apart. Perhaps I didn't like him because, in him, I saw a part of myself that I didn't like. Perhaps that was the part pursuing this escape plan.

We make fun of those we're most scared of becoming.

## LESSON 15

# AN ARGUMENT FOR GETTING INTO CARS WITH STRANGERS

Some people talk about the power of intent. They say that if you set your mind on something you really want, it will come to you. This is often misinterpreted as a rationalization for laziness, because it's a lot easier to lie in bed and dream than to go out and work.

Personally, I believe in the power of the odds. If you interact with enough people, and you look for clues leading to what you want every time, you'll eventually find someone or something that can help you.

This was what had happened with the conspiracy theorist at Alcor. And it was how I found my first real ally in this heretical quest: Spencer Booth.

He picked me up at JFK International Airport in a black Mercedes, which, like him, was tasteful but not flashy. Pale-skinned, with large red lips, ears that thrust violently out of his head, and a fleshy nose that seemed designed for a bigger face, Spencer didn't look like a billionaire. He reminded me more of an albino Mr. Potato Head. It was only when he spoke that his cheeks filled with color and his eyes with light, and it became clear that he was a man to be respected and reckoned with.

Road trips with strangers are long. Especially when they involve New York traffic. Along the way, I was indoctrinated into the world of Spencer Booth. He was from a world I'd never visited or even imagined before. It was the "B world," as he put it, full of "B people." B, I gathered, stood for a billion dollars. And B people, most of them big businessmen, were worth at least that much.

Spencer had just made a B after selling his latest technology company. Bored in the downtime afterward, he'd contacted me after reading my books because he thought I should start my own business. So he invited me to the house he was renting in the Hamptons to discuss it.

I took him up on the offer not because I actually wanted to be a businessman, but because I'd never been to the Hamptons before. I didn't have those kinds of friends.

Spencer pressed play on the CD player in his car and the deep, somber voice of one of my favorite songwriters, Leonard Cohen, prophesied through the crystal-clear sound system: "I've seen the future, brother: it is murder."

"The way to take over an industry," Spencer was saying over the music, beginning my first business lesson of the weekend, "is not to fix the current model, but to completely destroy it and replace it with a model you know is better."

"So let me ask you something, since you're a businessman." I didn't know why exactly—perhaps it was the music, perhaps it was a plea to the gods of the odds—but I felt like Spencer might understand what had been on my mind since Craig's terrorism speech at my book party. "Don't you think that's exactly what the terrorists want to do to America? They don't just want to destroy the country. They want to destroy the entire model."

"I do." Spencer turned onto the Sunrise Highway and sat silently in traffic. He glanced at me, scanning my face for something he could trust, then turned back to the road. "That's why," he continued slowly, nodding, deciding, "I hired a lawyer to help me get a second passport."

As soon as he said the word *passport,* my face exploded with color. I was excited, nervous, and, mostly, relieved. From the shift of energy in the car, it was clear this was no longer a business trip. I'd met a fellow runner.

It suddenly felt as if a weekend would not be long enough to talk about everything we needed to.

There is a theory called memetics, which suggests that ideas move through culture much like viruses. Thanks to the catalysts of 9/11 and Katrina, the escapist meme had clearly spread from the minds of fringe extremists to early adopters in mainstream society. It was only a matter of time, I began to worry, until countries further tightened their immigration policies because of the large numbers of Americans leaving, like their forefathers, for somewhere safer, more prosperous, and more free.

"What have you found out?" I asked Spencer. "So far, the only thing I know about is this organization called the Sovereign Society."

After returning home from Alcor, I'd found the Sovereign Society website. I'd actually stumbled across it earlier in my search but assumed it was just another scam to sell foreign real estate, international currencies, fraudulent passports, and expensive consultations.

This time, however, I noticed that the society was holding its first-ever Offshore Advantage Seminar in Mexico the following month. So I'd signed up in the hope of finding a community of like-minded escape artists who were further along in their quest for a safe haven.

"I've never heard of them," he replied. "I've been working with a lawyer, Holland Wright. When you get back to L.A., call him and tell him you know me. He'll take care of you."

Between the Sovereign Society and Spencer, I was no longer fumbling in the dark for an emergency exit. Suddenly, I had options.

Spencer parked in a gravel driveway outside his rental house and ushered me inside. Considering that he was sharing the place with one of the owners of a media empire, the vice president of a large mortgage company, a major hedge fund manager, and a venture capitalist from one of New York's wealthiest families, it was very un-opulent. Just a two-story white wooden building with five sparsely furnished bedrooms and a communal walkway that led over a ravine to the sea.

"So what country are you going to get your passport from?" I asked Spencer as he showed me to my bedroom. I had so many questions I wanted to ask, and so far he was the first person I'd met with any answers.

"I was thinking about the European Union."

"But I doubt Europe is going to be safe if there's another world war." I had discounted Europe immediately because anything America was involved in, Europe would surely be swept into as well.

"That's not the point. There are twenty-five countries in the European Union. And those countries possess territories all over the world. So with an EU passport, you can get somewhere safe from just about anywhere." His potato head contained so much knowledge. That's why he was rich and I wasn't.

"So what does it take to get an EU passport?"

In response, he brought me to the balcony, where a short man with a round, boyish face was talking on a cell phone.

"Neil, I want you to meet Adam," Spencer said. Adam was the venture capitalist. "He just got his Austrian citizenship this week."

I'd definitely stumbled into the right place.

## LESSON 16

# BIRTH CONTROL TIPS FROM BILLIONAIRES

While we waited for Adam to finish setting up a date on the phone, Spencer turned to me. "If you want female companionship this weekend, I can bring some up for you." He smiled cryptically, as if he were testing me. "I know these two breathtaking Russian girls who will really take care of you."

"Sounds kind of shady," I replied. A young Hispanic cleaning lady with Mrs. Kaufman–sized breasts mopped the floor begrudgingly in the living room. Her jaw was clenched, and her eyes shone with resentment at her humiliating task and those who paid her so little for it.

"They're not prostitutes or anything. They're just looking for wealthy guys to marry. That's pretty much their entire purpose in life. If I tell them you're rich and let you use my black card, you can have whichever one you want."

"But if they want to get married, why would they sleep with me on the first date?"

"Because it's going to be the best sex you've ever had. They know what they're doing. Just watch your condoms. They'll poke holes in them to get pregnant."

"That's one of the most devious things I've ever heard." No wonder Spencer was so paranoid.

Adam was still on the phone, complaining about his parents. It seemed strange to hear a man well into his thirties still calling them "Mommy" and "Daddy." On the end table next to him, I noticed a book, *The Rise and Fall of the Great Powers* by Paul Kennedy, and made a mental note to buy it.

"You don't even know the half of it," Spencer continued. "When I sold my company, I had to get my number changed. These kinds of girls check the newspapers for financial transactions, then wait outside the apartment of someone who's just made a lot of money hoping to meet them."

"Do they ever succeed?"

"The guys in this house would never marry one of those girls. What they're looking for is a very specific genetic stock. The girl needs to be beautiful, intelligent, and from a respectable family, with no history of hereditary illnesses."

"So they pretty much approach marriages like they approach their businesses?"

"That's why they're successful."

Rather than freezing themselves like Craig, these billionaires strove to extend their lives by creating new, improved versions of themselves. They didn't believe in children; they believed in a legacy. We were all, in our own way, running from death.

As Pulitzer Prize–winning psychiatrist Ernest Becker explained it in his book *The Denial of Death*, because we fear our own obliteration, we give our life purpose by embarking on an "immortality project" that will outlast us—whether it be our work, our children, the way we affect others, a good seat in the afterlife, or, in Craig's case, hope.

Adam hung up, then promptly called his parents. While we waited, Spencer suggested devising a plan to select my second citizenship. He loved plans—which was fine with me, because I'd always been bad at them. It was hard enough for me just to leave the house without forgetting my keys or wallet.

As if on cue, the maid started vacuuming loudly in front of us while Spencer took a pad of paper, set it down on the coffee table, and numbered it from one to five. By sorting through both practical necessities and personal preferences, we eventually selected five criteria required of the host country:

1. Must have a credible passport providing a wide network of visa-free travel.
2. Must be politically and regionally stable, with a low crime rate.
3. Must not significantly increase the tax liability of Americans living there.
4. Must not require more than two years of residency for citizenship.
5. Preferably in a warm climate with beaches.

Admittedly, the last criterion was more personal than political.

Adam soon joined us. He reminded me of a thinner, better-groomed version of Ignatius J. Reilly, the overweight mama's boy with an unwarranted superiority complex in the novel *A Confederacy of Dunces*.

As he sat down, he glanced up in irritation at the maid. She met his gaze coldly and continued vacuuming the same spot. She was obedient when under observation but seemed like she'd drop poison in his coffee the moment he turned his back on her.

"Neil wants to know how you got your passport," Spencer prompted him.

"My passport?" Adam asked. "Anyone can get one"—a smug smile spread across his face, like that of the only kid in the play-

ground with a chocolate bar—"if they invest over a million dollars in the country."

"In Austria?"

"It's the only place in the European Union that will give you citizenship for making an economic investment. I started a venture capital business there and hired a bunch of Austrians. It took forever to get approved. I had to go to the highest level of government."

I didn't have that kind of money to invest, let alone those kinds of connections. It looked like the European Union was for the B boys. "How long did it take to get it?"

"I started trying as soon as Bush was reelected."

Suddenly, I didn't feel so foolish. If the smartest, richest guys in the country were doing this, then clearly I wasn't paranoid. I was just ahead of the curve. St. Slim Jim, prophet of passports.

"A bunch of other B people are doing similar things right now," Spencer told me. "Do you know the Walton family? They own Wal-Mart. They just built an underground bunker near their home in Arkansas. They even have a helipad in case they need to evacuate."

"That's crazy." It was amazing how little difference there was between the billionaires and the cult leaders.

"Actually, it isn't. If something does happen in America, it may be difficult to get out," Spencer replied slowly, as if confiding a secret he hadn't meant to tell me. "So we're taking flight lessons in a few months."

Spencer didn't start any business without researching every minute detail, then drawing up a schedule that looked forward at least ten years, included all expenses, and had contingency options for every possible obstacle, including death. His escape-from-America plan was no different.

"I'm not taking any chances with my family," he continued. "I just bought them guns, in case we have to shoot our way to the airport."

Coming from the mouth of a respected businessman—especially after watching the Fliesian looting in New Orleans—these extreme preparations were actually beginning to sound reasonable. Except for the guns. I couldn't imagine killing anyone. Unfortunately, as Bettie the goat knows, that lofty ideal of mine would fall by the wayside as the country continued its downhill slide.

Instead of inviting the Russian gold diggers over, Adam and I spent the rest of the night drinking wine and debating his escape route with his housemates.

"I'm executing a ten-year plan to make sure everything that could go wrong is protected against," Spencer said, his face glowing with the pride of feeling one step ahead of everyone else. "It's about creating revenue sources and residences in multiple locations, so that if you have to flee one country, your daily existence won't change."

"Do you know the problem with your plans?" challenged the mortgage company vice president, Howard, a pasty, heavyset man with an incongruent passion for mountain biking. "You don't know what's going to happen, so there's no way to prepare. If there's never a disaster, then you're just wasting your time like all those people who built bomb shelters in the sixties. And if there is a disaster and someone drops a nuke, it'll probably be in New York and you'll get vaporized. So why not just relax and enjoy life, instead of worrying about everything that can go wrong?"

"Because this makes economic sense," Spencer replied coolly. "Do you have insurance for fires, theft, and illness?"

"Of course," Howard replied.

"Which means you agree that a certain portion of your income should be allocated to protect you against catastrophes, even if they're low-probability events." Spencer took a small, prudent sip of wine, then continued. "Because that's all this is: an insurance policy. When you think something will never happen to you, that's usually when it happens."

LESSON 17

# THE BRITNEY SPEARS SCHOOL OF PREGNANT SURFING

The next morning, as Spencer and I sat on the patio, a tall, sullen man let himself into the house and stormed to the kitchen. He acknowledged no one, and no one acknowledged him. Evidently, he was the cook. As he grabbed pans and utensils with quick, angry movements, the maid moved through the house at a defiantly slow pace, collecting dirty linens.

"We need to make you a ten-year plan," Spencer said, beginning my next business lesson. "Before I started my company, I sat down and planned everything. It took a decade to execute, but in that time, nothing happened that wasn't part of the plan. So to start, let's make a list right now of the projects you're working on."

I gave him a thumbnail sketch of future books. Most recently, I'd received a call out of the blue from Britney Spears. I had interviewed her for an article and we'd exchanged phone numbers. Almost a year later, she had called and invited me to her house in Malibu.

"Do you think I can try surfing with you like this?" she asked with touching naïveté when I arrived and told her she lived across the highway from my favorite surf spot. She was at least five months pregnant with her second child.

She had called me over, she explained as we sat down on her couch, because after I'd interviewed her, she'd researched me and seen my books. Now she wanted me to write her life story.

"I want it to be something like this," she said. In her hand was a copy of Goldie Hawn's *A Lotus Grows in the Mud*. In her other hand was a crumpled piece of white paper. On it, she had written notes from her life: about cheating on Justin Timberlake, about resenting her parents' control, about sexually acting out.

"So are you going to write it?" Spencer asked after I told him the story.

"I think it'll be a good book if she's going to be that honest," I said. "But afterward, things got really strange. Her manager, her agent, and her lawyer were all fighting for control of her. Some of them wanted to do the book; others didn't. She started calling me, but for some reason they took away her phone and it became impossible to get in touch with her."

Spencer went silent and thought for a while. "Okay," he said, "so we have the Britney situation, the passport situation, and the ten-year plan. What if we put them all together?" I loved the way his potato mind worked. I needed more people like this in my life.

"I don't know if I want to put them together. The Britney book is probably a bad idea. Her people would never let the truth come out."

I noticed the cook talking to the maid in a hushed French accent. It was the first time I'd seen either of them smile. As soon as they realized I was looking, they separated and went about their business. It felt like a class war was brewing in the house.

"You know what would be great?" Spencer went on. "Find a killer place in some third-world paradise and spend a relatively insignificant amount of money building it into a compound." The word *compound* filled my head with images of the House of

Yahweh. I couldn't believe I was seriously contemplating this. "Then start a publishing company. If someone like Britney wants to work with you, you can bring her and her entire entourage there. Make sure they're taken care of. Every day her friends can play while you work. The property will pay for itself, because not only will you be the writer everyone wants to go to, but you can start publishing books by other celebrities as well."

"But I don't have the kind of money you do."

As the cook chopped vegetables with no affection for the task, I thought about how little it would take for the have-nots to revolt against the haves. When those who feel like they're being treated unjustly have a target for their resentment, they become coiled springs pressed toward the ground. Whether you add more pressure or take it away, the result is the same: they will rise against you.

"You don't need money," Spencer responded. "No intelligent person spends his own money. Have your publisher fund it as a joint venture. They can bring other writers and clients there too."

Suddenly, leaving the country felt less like an escape from the rat race and more like an opportunity to run it better and faster than the other rats.

That night, after a house dinner during which the B boys gossiped about a friend who almost brought the New York Stock Exchange to a standstill, Spencer and I sat in the living room and watched television. On the news, North Korea, upset over sanctions imposed by the United Nations after the country conducted a nuclear test, was threatening to "mercilessly" retaliate.

Spencer and I didn't need to say a word to each other. We were both thinking the same thing.

## LESSON 18

# RAISING THE DEAD WITH GOVERNMENT PAPERWORK

Back in Los Angeles, I bought the book I'd seen at Spencer's house, *The Rise and Fall of the Great Powers* by Paul Kennedy, which, from the first page, was eerily prophetic.

Written in 1987, the book meticulously traces the histories and economies of world superpowers of the last five hundred years. Its thesis, born of an endless array of historical examples, is that the collapse of every superpower in modern history has been due not just to lengthy fighting by its armed forces, but to its interests expanding internationally while its economy weakens domestically. In other words, empires collapse when they stretch themselves too thin.

Furthermore, Kennedy explains, "Great Powers in relative decline instinctively respond by spending more on 'security,' and thereby divert potential resources from 'investment' and compound their long-term dilemma."

Though he doesn't discuss America for most of the book, every word and every example has parallels in the United States today. For example, he writes of the declining Spanish empire in the sixteenth century, "Spain resembled a large bear in the pit: more powerful than any of the dogs attacking it, but never able to deal with all of its opponents and growing gradually exhausted in the process."

Similarly, two hundred years later, the British, "like all other civilizations at the top of the wheel of fortune . . . could believe that their position was both 'natural' and destined to continue. And just like all those other civilizations, they were in for a rude shock."

As each successive empire fell on the same economic sword, Hegel's words kept echoing in my head: "What experience and history teach is this—that peoples and governments have never learned anything from history."

By the time I finished the book, it seemed clear that whether or not there was another devastating terrorist attack in America, the end of the empire as we know it was approaching. The only remaining questions were: Would it simply become a smaller player, like Spain and Britain after their zenith, or disappear altogether, like Rome, Austria-Hungary, and the Soviet Union? And what world power would eventually take its place: China, the European Union, maybe India?

Either way, I was now convinced that, as surely as one tucks money into a retirement plan, I needed to build a backup life offshore immediately—and maybe even take Spencer's advice and look into other ways to protect myself. In a time when nuclear, chemical, and biological weapon technology is for sale to the highest bidder, especially since the collapse of the Soviet Union, the fall of a great empire today could be a lot messier than in the past.

After returning home from the Hamptons, I'd also called the lawyer Spencer had recommended, Holland Wright.

"Your clients Spencer Booth and Adam McCormick recommended I speak to you," I told him, my voice slightly nervous, as it was every time I knew I'd have to say the next sentence aloud.

"I'm looking for a second citizenship and wanted to see if I could work with you."

"That's not something I do," he replied brusquely.

"It's okay." I assumed he was just trying to protect his clients' confidentiality. "I was with them in the Hamptons this weekend. They told me to call you."

"No, that's not something I do," he repeated. Maybe I wasn't B enough for him.

"But—"

"Is there anything else I can help you with?"

I hate lawyers.

"Thanks for your time." I hung up dejectedly.

I needed to find someone who understood the advantages and disadvantages of each country, the benefits and restrictions of each passport, the various legal shortcuts to obtaining quick citizenships. If necessary, I was even willing to consider getting married.

But no lawyer would help me.

Before leaving for the Sovereign Society conference, I researched one other option: a tombstone ID. Spencer had told me he'd considered getting one but changed his mind because it was too shady.

To obtain a tombstone ID from a foreign country, I'd need to fly there and go to a local cemetery. Then I'd have to find the grave of a boy who'd died between the ages of three and ten, write down his name and birth date, and take out a mailbox in his name.

After researching information about his parents on a genealogy website, I'd have to apply to the government for a birth certificate. With that certificate, I could get the country's equivalent of a social security number and card. And with both those documents—and perhaps, for extra security, a utility bill in that person's name or a fake student ID—I could get a driver's license.

Finally, with all those documents (and, in some countries, a local friend to vouch that I was a citizen), I could get a brand-new passport.

But in addition to being risky—especially since more municipalities were starting to store birth and death certificates in the same database—there was also something degenerate about having my face on a passport above the name of a dead child.

As I boarded a plane to Mexico, I hoped someone at the Sovereign Society conference would have the answers I was looking for. I was running out of legal options.

# HOW TO EVADE PURSUIT VEHICLES

## 1. MODIFY YOUR VEHICLE

PREPARE AHEAD OF TIME WITH, AT A MINIMUM, RUN-FLAT TIRES THAT WILL OPERATE AT HIGH SPEEDS WHEN PUNCTURED. IF POSSIBLE, ALSO ADD HIGH-QUALITY SHOCKS AND SPRINGS, BULLET-RESISTANT WINDOWS, STAINLESS-STEEL BRAKE LINES, A HEAVY-DUTY RADIATOR, AND DUAL RAM BUMPERS.

IF YOU WANT TO GET SERIOUS, ADD LAYERS OF KEVLAR ON THE CAR INTERIOR, BALLISTIC WRAP AROUND YOUR GAS TANK, A DUAL-BATTERY SYSTEM, AN ELECTRIC-SHOCK SYSTEM ON THE CAR EXTERIOR, AND STEEL PLATES (WITH GAPS FOR AIRFLOW) PROTECTING THE ENGINE. KEEP IN MIND THAT ANY ADDITIONAL WEIGHT WILL AFFECT THE CAR'S HANDLING.

## 2. STOP THE CHASE BEFORE IT HAPPENS

QUICKLY DISABLE UNOCCUPIED PURSUIT VEHICLES BY STICKING A KNIFE INTO THEIR TIRE SIDEWALLS OR SHATTERING THEIR FRONT WINDSHIELDS.

## 3. BLIND THE ENEMY

CARRY A HANDHELD SPOTLIGHT OR 500-PLUS-LUMEN FLASHLIGHT TO SHINE INTO THE EYES OF PURSUING DRIVERS. IDEALLY, INSTALL SPOTLIGHTS OR FLASHING STROBE LIGHTS ON YOUR VEHICLE.

## 4. FLATTEN YOUR OPPONENT

BUY CALTROPS, OR MAKE ONE BY HAMMERING HALF A DOZEN NAILS INTO A GOLF BALL SO THAT IT WILL LAND WITH A POINT IN THE AIR. DROP THEM INTO THE ROAD BEHIND YOU. IF YOU HAVE TIME, CONCEAL THEM INSIDE A STYROFOAM CUP.

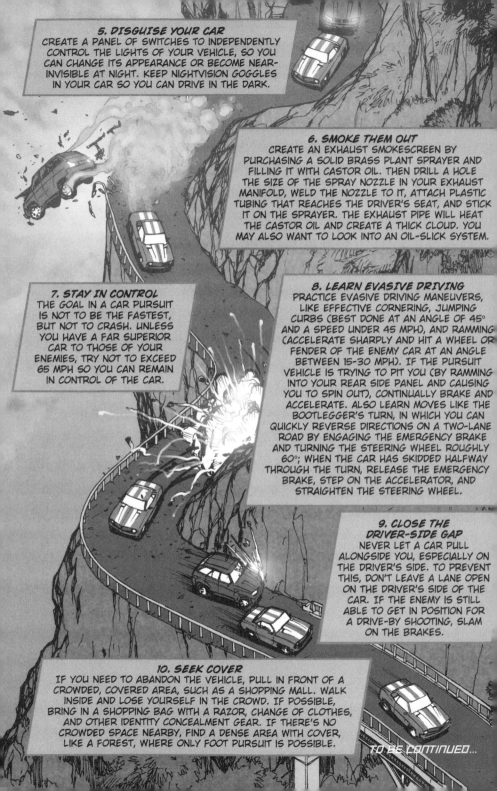

**5. DISGUISE YOUR CAR**
CREATE A PANEL OF SWITCHES TO INDEPENDENTLY CONTROL THE LIGHTS OF YOUR VEHICLE, SO YOU CAN CHANGE ITS APPEARANCE OR BECOME NEAR-INVISIBLE AT NIGHT. KEEP NIGHTVISION GOGGLES IN YOUR CAR SO YOU CAN DRIVE IN THE DARK.

**6. SMOKE THEM OUT**
CREATE AN EXHAUST SMOKESCREEN BY PURCHASING A SOLID BRASS PLANT SPRAYER AND FILLING IT WITH CASTOR OIL. THEN DRILL A HOLE THE SIZE OF THE SPRAY NOZZLE IN YOUR EXHAUST MANIFOLD, WELD THE NOZZLE TO IT, ATTACH PLASTIC TUBING THAT REACHES THE DRIVER'S SEAT, AND STICK IT ON THE SPRAYER. THE EXHAUST PIPE WILL HEAT THE CASTOR OIL AND CREATE A THICK CLOUD. YOU MAY ALSO WANT TO LOOK INTO AN OIL-SLICK SYSTEM.

**7. STAY IN CONTROL**
THE GOAL IN A CAR PURSUIT IS NOT TO BE THE FASTEST, BUT NOT TO CRASH. UNLESS YOU HAVE A FAR SUPERIOR CAR TO THOSE OF YOUR ENEMIES, TRY NOT TO EXCEED 65 MPH SO YOU CAN REMAIN IN CONTROL OF THE CAR.

**8. LEARN EVASIVE DRIVING**
PRACTICE EVASIVE DRIVING MANEUVERS, LIKE EFFECTIVE CORNERING, JUMPING CURBS (BEST DONE AT AN ANGLE OF 45° AND A SPEED UNDER 45 MPH), AND RAMMING (ACCELERATE SHARPLY AND HIT A WHEEL OR FENDER OF THE ENEMY CAR AT AN ANGLE BETWEEN 15-30 MPH). IF THE PURSUIT VEHICLE IS TRYING TO PIT YOU (BY RAMMING INTO YOUR REAR SIDE PANEL AND CAUSING YOU TO SPIN OUT), CONTINUALLY BRAKE AND ACCELERATE. ALSO LEARN MOVES LIKE THE BOOTLEGGER'S TURN, IN WHICH YOU CAN QUICKLY REVERSE DIRECTIONS ON A TWO-LANE ROAD BY ENGAGING THE EMERGENCY BRAKE AND TURNING THE STEERING WHEEL ROUGHLY 60°; WHEN THE CAR HAS SKIDDED HALFWAY THROUGH THE TURN, RELEASE THE EMERGENCY BRAKE, STEP ON THE ACCELERATOR, AND STRAIGHTEN THE STEERING WHEEL.

**9. CLOSE THE DRIVER-SIDE GAP**
NEVER LET A CAR PULL ALONGSIDE YOU, ESPECIALLY ON THE DRIVER'S SIDE. TO PREVENT THIS, DON'T LEAVE A LANE OPEN ON THE DRIVER'S SIDE OF THE CAR. IF THE ENEMY IS STILL ABLE TO GET IN POSITION FOR A DRIVE-BY SHOOTING, SLAM ON THE BRAKES.

**10. SEEK COVER**
IF YOU NEED TO ABANDON THE VEHICLE, PULL IN FRONT OF A CROWDED, COVERED AREA, SUCH AS A SHOPPING MALL. WALK INSIDE AND LOSE YOURSELF IN THE CROWD. IF POSSIBLE, BRING IN A SHOPPING BAG WITH A RAZOR, CHANGE OF CLOTHES, AND OTHER IDENTITY CONCEALMENT GEAR. IF THERE'S NO CROWDED SPACE NEARBY, FIND A DENSE AREA WITH COVER, LIKE A FOREST, WHERE ONLY FOOT PURSUIT IS POSSIBLE.

TO BE CONTINUED...

LESSON 19

# HOW TO PROTECT YOURSELF FROM INFLATION, HACKERS, AND CELINE DION

A few words of advice: never go to a conference of paranoid people looking for information.

I've spent fifteen years as a reporter. I've accidentally found myself swept up in riots, coups, and Kenny G concerts, and always gotten the story. But the Sovereign Society Offshore Advantage Seminar was no place for the casual question.

The wide hallways outside the conference rooms of the Sheraton Hotel in Puerto Vallarta, Mexico, were packed with middle-aged men and women. They weren't well dressed, but they weren't badly dressed either. They weren't attractive, nor were they ugly. They didn't appear to be rich, but they didn't seem poor.

What bound them together was that they all appeared inconspicuous. Unlike me, they weren't the type of people selected for random searches by airport security. Yet there was one thing that separated them from the rest of the guests at the Sheraton: they were alone. None appeared to be with spouses, children, or families. Their mission, and perhaps their life, was a solitary one.

As was mine. I hadn't yet told my family or friends about what I was doing. As far as they knew, I'd gone to Mexico to surf, drink, and meet girls. Which definitely would have been more fun.

I looked around the hallway—at the potbellied man in the Hawaiian shirt, the greasy-haired woman in the blue business suit, the unshaven man in the Panama hat. And I wondered if they'd already escaped, and where and when and why and how it went. But I would never know. Outside of a few pleasantries, none of them would talk to me.

I took a seat in the back of the conference room, hoping to glean something from the presentations. A man in a business suit with black hair cut over big ears was speaking onstage. His name was Rich Checkan. His company was Asset Strategies International.

He said that when a country's trade deficit is greater than 5 percent of its gross domestic product, the currency typically falls 20 to 40 percent.

He said that the U.S. deficit was at 5.7 percent of its gross domestic product, making a substantial dollar devaluation practically unavoidable.

He said that people trying to escape wars had been able to barter diamonds and gold to get airlifted out of the country.

He said that to protect ourselves, we should invest in foreign currencies and precious stones and metals, particularly gold.

He said he happened to offer just such a service.

None of the presentations that day were about obtaining a passport and fleeing America. The world of the Sovereign Society was much more involved than that. Thomas Fischer of Jyske Bank in Denmark urged attendees to keep their money offshore, where it wouldn't get confiscated by the government or lost in a lawsuit. Mark Seaton of Armor Technologies tried to sell computer privacy to prevent hackers from accessing private data. Chatzky and Associates urged attendees to open foreign trusts and LLCs to protect their assets.

I looked over the schedule for the second day. Most were also

about keeping money safe from high taxes, lawsuits, asset seizures, government regulations, and inflation.

The paranoia here was different from mine. The members of the Sovereign Society were suspicious of the United States government. And while the events of the last few years compelled me to agree with them, I was equally worried about the suspicion other countries had of the U.S. government—and the preemptive or retaliatory measures they might take against it. In short, they wanted to save their money. I wanted to save my skin.

By the end of the day, I still hadn't made a friend or been in a discussion longer than three minutes. Some people seemed to know each other, but whenever I tried to join their conversations, they ignored me. By this point, I was pretty sure I was the most suspicious-looking person there.

Though paranoia is often used as a derogatory term, the truth is that it's a survival instinct. If you think your postman is stealing your checks or your nurse is poisoning your food, even if it's not true, the accusation is rooted in an innate desire for self-preservation. When a cat perks up its ears or hides under a bed when a completely harmless stranger approaches, it's the exact same response, developed through millennia spent living in the wild, where the unknown was a threat.

And because the future is unknown, no matter how good or bad things may be today, it will always be a threat. So, ultimately, the sole arbiter of what's paranoia and what's common sense is what happens tomorrow. You're only paranoid if you're wrong. If you're right, you're a prophet. St. Slim Jim, patron saint of offshore bank accounts and paranoid cats.

Though the presenters were easy to mock, the next day I found myself picking up a flyer for Jyske Bank in Denmark from the conference information booth. If I was going to set up a new citizenship offshore, then I might as well put some money there to protect my savings against hyperinflation, bank failures, and lawsuits. Just in Case.

By writing books and articles, I was exposing myself to litigation and violating one of the key principles of escape artists everywhere: privacy. The less people know about you, the less open you are to attack. I'd had many lawyers—including attorneys for Celine Dion and Michael Jackson—threaten me while I was at the *Times*.

My friend Craig, who'd recently received his silver cryonics wrist bracelet in the mail, had been sued by a former business partner. The litigation had wiped his bank account clean, he still had $150,000 in legal bills to pay, and the case hadn't even gone to court yet. Then there was my friend Frank, an Internet marketer who used the wrong disclaimers on his website. He was consequently accused of running a pyramid scheme by the Federal Trade Commission, which seized every penny he had without filing a single criminal charge.

I wasn't rich by any means, but it would still be cool if my net worth was zero. If I didn't own anything, then nothing could ever be taken away from me.

The speakers had clearly gotten to me.

We make fun of those we're most scared of becoming.

As long as I was gathering intelligence, I grabbed a USB stick with information about Armor Technologies on it as well. Considering that I was about to take nearly every step the U.S. government and the IRS considered suspicious, it was time to protect my computer from prying eyes.

As I turned to walk away, I noticed a short woman shuffle into the information booth. Her name tag identified her as Erika Nolan, president of the Sovereign Society. Maybe she would talk to me.

"I need some help," I told her. "I want to get a second citizen-

ship somewhere. And I was hoping you knew the easiest way to do this."

She hesitated, as if weighing her options for escape, then relented. "What kind of places are you considering?"

"I'm not sure. But it needs to be somewhere safe and with a credible passport that allows visa-free travel."

"Well, I'm not an expert in all the different immigration laws." As soon as she said this, I wondered if I'd ever find what I was seeking. "But tomorrow there are two speakers you may want to see. One is Wendell Lawrence, who's here from St. Kitts. He used to be an ambassador there. The other is Ramsés Owens, who's going to discuss economic investment programs in Panama."

"That's great. But what I'm really looking for is an expert or a lawyer who knows the rules for every country and can discuss the pros and cons of each one."

She nodded as if she understood and then lifted the tablecloth, reached underneath, and pulled a big blue book from a large stack. When I saw the cover, I knew I had to own it:

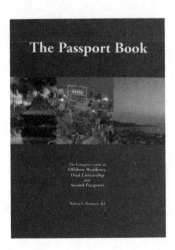

This was my ticket out.

## LESSON 20

# FIVE STEPS TO A TAX-FREE LIFE

That night, as I ate alone and friendless on the hotel restaurant patio, I cracked open *The Passport Book*.

Travel documents have been around for centuries. At first, they were letters from an authority (a king, a priest, a pharaoh, a landowner) allowing the bearer to travel unobstructed. While some argue that travel documents provide freedom, others maintain they've historically been used for control. By selectively granting and withholding them, governments have been able to regulate who enters and leaves their country. In the United States, outside of brief periods during previous wars, it wasn't until 1918, near the end of World War I, that the government began requiring everyone crossing the border to have a passport.

"Is that *The Passport Book*?" Someone was actually talking to me.

I looked up to see a thin man in baggy jeans and a short-sleeved green shirt. "Be careful with that," he warned. "The information goes out-of-date pretty quickly."

Without hesitation, I blurted the question I'd wanted to ask every single person at the conference. "Do you already have a second passport?"

"I was born in the UK, but I moved to New Zealand when I was ten, so I already have two passports. But actually"—a proud smile spread over his lips—"I don't consider myself a citizen of any country."

I knew what this meant, because I had already come across it in my research. I had just met my first PT.

"So you're basically PT?" I asked.

"Since 2000," he said, beaming. I couldn't believe he was opening up to me.

Half a century ago, after serving in World War II, a newspaper publisher named Harry Schultz returned to America and was disappointed to find a nation of violent crime, high taxes, and frivolous lawsuits. So he sold his thirteen newspapers and decided to become not a citizen of America, but a citizen of the world. The name he gave to this idea was PT. The letters don't stand for any two specific words, but they're most often defined as *perpetual tourist* or *permanent traveler*.

The idea of PT is that, just as we shop at different stores in a mall to find various items we want, we can also shop in different countries to find the lifestyles, governments, careers, people, tax rates, and cultures that best suit us.

So why stay in America just because you were born here? There's a great big world out there with a lot to offer. Just as every child must eventually grow up and leave home, just as our ancestors left the Old World in search of the new, just as the hero of every great myth journeys outside the familiar, so too must we venture outside our small reality and, rather than simply believing we're living the best in the best of all possible countries, find out for ourselves.

During my months of fruitless searching for a passport mentor, I had e-mailed Schultz, who was in his eighties. He said he would answer my questions, but, because his eyesight was failing, asked that I fax them to him in very large type. A few months later, I received a response that was several pages long. Unfortunately, he must have answered someone else's fax, because he'd responded to my questions about PT and citizenships with predictions for the gold, stock, and housing markets.

Although I didn't get the advice I needed, I did receive some much-appreciated encouragement. "Govts cant save U," he wrote at the end of the e-mail. "They aren't on your side. U have to save yourself. Row your boat. Be your own country. Wave no flag but your own."

I hoped I would still be that passionate when I was his age. Life is conveyed not just through a heartbeat and brain wave activity, but through an immeasurable spark that animates our faces, our conversations, our being. That spark may be the pursuit of love, success, excitement, validation, connection, happiness, learning, God, or freedom. Its fuel is hope. Without a belief in a better future, it dies. And, aside from an early death, my greatest fear is one day losing that spark, whether through gradual disillusionment or sudden calamity.

The PT at the hotel introduced himself as Greg and invited me to join him at his table. He had the spark. On a scale of one to ten, it was an eight. "I sold everything, paid my tax bill, and told the New Zealand government I was leaving the country," he said, his cheeks flushed with pride.

I was surprised to hear this, because New Zealand seemed like the safest English-speaking country left. "What made you leave?"

"I found myself in a bad situation. I'd lost all my money, including my house. So my son and I moved into a friend's home. I was about to start a day job teaching computer programming, but then I realized that the system was skewed against me."

"What do you mean?"

"Like America, New Zealand has a progressive tax system. The more you earn, the higher your tax rate is. I realized that my

dreams of getting ahead were up against this system that was going to penalize me for being successful. So I decided I wasn't going to stay there to be milked by the government. And I followed the PT course to remove myself legally from the system."

Greg said he'd discovered the concept of PT after embarking on a libertarian reading jag. Ayn Rand's classic tome on capitalism and individualism, *Atlas Shrugged,* had led him to former Libertarian Party presidential candidate Harry Browne's influential *How I Found Freedom in an Unfree World.* From there, he'd picked up *Sic Itur Ad Astra* (This Is the Way to the Stars), a workbook by the astrophysicist Andrew J. Galambos on how to build a society based on personal freedom. And this had led him to the writings of a man known by the pseudonym W. G. Hill.

"Do you have to be a libertarian to be PT?" I asked. Where libertarians believe the best government is the one that governs least, I'd always believed the best government is the one that governs best.

"Not at all," he replied, and then opened his computer and showed me a scanned copy of Hill's book *PT.* In the introduction, Hill claims he was a millionaire whose conspicuous display of wealth led to problems from ex-wives, tax auditors, lawyers, and employees, all of whom wanted a piece of the pie. He was eventually imprisoned for fraud and his assets were seized. While trying to put his life back together, he ran across a pamphlet by Harry Schultz.

After reading it, Hill became excited by the prospect of, in his words, "a stress-free, healthy, prosperous life not limited by government interference, the threat of nuclear war, the reality of food and water contamination, litigation, domestic conflicts, taxation, persecution or harassment."

The way to break free of nationality, according to Schultz's pamphlet, was to follow the three-flag system. The three flags consist of having a second passport, a safe location for your assets in another country, and a legal address in a tax haven. To these, Hill added a fourth and fifth flag: an additional country as a business base and a number of what he called "playground countries" in which to spend leisure time.

"Seems complicated," I told Greg. I couldn't wrap my head around the concept, perhaps because I wasn't really business savvy. My understanding of money was still primitive: You do work, you get paid for it, and you do your best to put some of that money in a bank.

"The core of PT," Greg explained patiently, "is that if you're prepared to decouple from your home country—which is quite an emotional thing for most people—and not spend more than a hundred and eighty days in more than one place, unless it's a tax haven, you can legally step outside of the obligation to pay income tax. The various flags are designed as a practical way of doing that."

Though I have my own problems with taxes—they're too high, they're painful to pay, and too much of the money goes to defense contractors rather than domestic improvements—I understand their necessity. Unlike Thoreau in *Civil Disobedience,* I'm not going to just stop paying taxes and go to jail for it. If we weren't giving the government money to protect us, then there would be other people demanding protection money instead. So why not just pay the government, which is at least accountable to a degree—unlike, say, the Mafia?

"It's a lot of effort to escape from taxes," I told him. "Couldn't the same amount of work just be put into making more money to offset your tax liability?"

"I see." I hoped I hadn't already offended the only person who'd been nice to me here. "For me, I guess, the main advantage to being outside the system is not the money. It's the inde-

pendence and the freedom from bureaucracy. The first thing I did after I left New Zealand is I flew to Australia and went for a walk. I had the most amazing sense of freedom." His cheeks filled with color again. Evidently, he hadn't taken my skepticism personally. "I was really on cloud nine. I felt like I had been released from a sense of claustrophobia."

That feeling I understood.

"You should get in touch with a guy called Grandpa in Monaco," Greg suggested after I blitzed him with follow-up questions. "He's the archetypal PT. He might be able to help you."

Before retiring to my room, I asked him why everyone else at the conference was so reluctant to talk. "You have to be careful in this world," he explained. "A lot of people get into PT because they're hiding from the law. Others get into it so they can scam folks who are desperate for passports or anonymity. Even the honest people are, by nature, private and distrustful. So it's a world of people trying to stay in the shadows. Good luck trying to get anyone else to talk."

"Why were you so nice to me then?"

"Because I found the sixth flag."

"What's that?"

"Freedom from fear."

## LESSON 21

# WHY KNOCKING UP A BRAZILIAN WOMAN CAN SAVE YOUR LIFE

A s soon as I returned to my hotel room, I tore *The Passport Book* open again and skimmed the nationalization requirements for each country, hoping for an answer.

According to the author, several countries have loopholes that allow easier citizenships based on technicalities—for example, conceiving a child with a Brazilian woman, having a parent or grandparent who was a citizen of Ireland, proving a parent or grandparent was a German refugee during World War II, or having a Greek citizen for a father.

Unfortunately, I had no relatives who were Irish or Greek, none of my German relatives were refugees, and I hadn't impregnated any Brazilian women lately.

The easiest option was Israel, which grants citizenships after ninety days of residency not just to all Jews but to anyone who converts to Judaism. However, not only would Israel be the worst place to live if a major war broke out in the Middle East, but military service is compulsory for both men and women there.

When I finished *The Passport Book,* I researched W. G. Hill online, then ordered electronic versions of three of his books.

To my disappointment, I discovered that the expression "In this world, nothing is certain but death and taxes" is acutely true for Americans, making it nearly impossible to become a PT like Greg.

After a 1994 *Forbes* article titled "The New Refugees" exposed a cabal of billionaires like Kenneth Dart (whose father invented the Styrofoam coffee cup) and John Dorrance III (whose father invented condensed soup) that had expatriated to avoid taxes, Congress passed a law obligating most Americans who renounce their citizenship to continue to pay taxes for another ten years, regardless of whether they reenter the country or not.

Another loophole allowed Americans to obtain a second citizenship, become a consul in their new home country, and then claim diplomatic immunity from U.S. taxes. But the government had made that impossible as well. In his book, Hill suggested a few workarounds, but they entailed never coming back to America—or returning only under an alias. And I wasn't ready to make that sacrifice.

All this research into escaping America was inadvertently making me appreciate it more. Not only was I unwilling to forsake the country permanently, but there wasn't anywhere else I was willing to live full-time.

As the sun rose outside my hotel window, I fell asleep discouraged, my head flooded with inconclusive data. I felt like I was living in a house with no doors. I was happy there, but if a fire broke out, I'd be burned alive. I needed an emergency exit.

When I walked into the conference room late the next morning, a speaker was warning attendees that information on every bank transaction of more than $10,000 is sent to the U.S. Treasury to track potential financial crimes. I wondered how many people in the audience were actually wealthy enough to make all these asset protection and privacy efforts worth the effort. They seemed to be richer in fear than actual money.

In the back row sat my first hope for escape: a bald, well-dressed Panamanian named Ramsés Owens, Esq. I approached and asked him about the requirements for citizenship in his country. He informed me that it would take five years to get a passport, though I could get an investor visa to obtain instant residency by investing $100,000 in a Panamanian business.

I thanked him and moved on. I was hitting dead ends everywhere. Walking away from the seminar empty-handed would leave me with little option besides a tombstone passport.

On the other side of the room, Wendell Lawrence filled a folding chair. He was a big man, reminiscent of a slightly rounder and friendlier-looking Forest Whitaker. He wore a gray button-down shirt tucked into beige trousers pulled high on his waist. Yet he wore the outfit so cleanly and properly that it seemed not like he'd pulled his pants up too high, but as if everyone else had let theirs drop too low.

"Erika Nolan suggested talking to you," I said, meekly introducing myself.

"Sit down," he replied in a deep Caribbean accent.

I took a seat in the shadow of his bulk. Though he loomed over me, he wasn't intimidating. He was inviting. I felt safe, perhaps because of the way his eyes glittered with almost paternal friendliness.

"So what can I do for you?"

"I wanted to see what steps I'd need to take to become a citizen of St. Kitts."

"Ah," he said, clapping his hands together. "You'll love St. Kitts. It's a small and beautiful island, with friendly people and real Caribbean culture. You should come for Carnival this year."

He was like a living sales brochure. A very convincing one. He turned his seat to face mine and continued. "If you're interested in living in St. Kitts, there are two ways to get citizenship. The fastest way is three months."

It seemed too good to be true. There had to be a catch.

"As you may know"—I didn't, of course—"the sugarcane industry on the island collapsed. This year, the last sugarcane plantation closed. So"—here was the catch—"by investing in industries that will provide jobs to out-of-work sugarcane employees, you're granted citizenship instantly. The minimum investment used to be a hundred thousand dollars. But that just went up to two hundred thousand dollars."

"How long does it take to get that investment back?"

Wendell let out a long laugh. "Oh, no. You don't get the money back."

"So it's more like a donation."

"There's also another way," he continued. "This is by investing in real estate. You must buy a piece of approved property that costs over three hundred and fifty thousand dollars."

"Can I live in it and sell it when I want?"

"It's your home, but you have to keep it for at least five years to retain your citizenship."

That sounded like a better deal—especially since wherever I went, I'd need a place to live anyway. To pay for it, I could always get a loan. This was back in the good old days, when mortgage companies gave interest-only loans to practically anyone who asked, no down payment necessary.

"How long does the citizenship process take?"

"It takes longer than the investment program. About six months."

I wanted to punch my fist jubilantly into the air, but I tried to contain my excitement—not just out of decorum, but out of caution. I needed to talk to Wendell a little more and see if he and this program were for real.

With pride, he told me about the island's high literacy rate. Its free medical care for children, senior citizens, and people with chronic diseases. Its public school system, which provides free textbooks and lunch for students, in addition to computer labs open to everyone on the island. And its status as a financial haven, with no income tax for citizens. My new PT friend, Greg, would approve.

Either Wendell was a genuinely good person with a heartfelt passion for his native country, or he was the best con artist in the room.

That night, I called Spencer and excitedly told him about the St. Kitts real-estate program.

"That seems like it could work," he replied. He sounded pleased, though not as excited as I'd hoped, especially considering he hadn't found any second citizenship options for himself yet. "It would be great if we both had places there. I'll have my lawyer look into it."

"Your lawyer's an asshole."

"All lawyers are assholes."

Between the conference and my conversation with Greg, I had now fully come around to the idea that if anything went wrong in America, it would be nice to get out with not just my life but my bank account. So I asked Spencer what he'd done with his money.

"I flew out to Switzerland and met with the different banks," he replied. "I wanted to ask about their history during World War II and find out if they gave back the gold deposits they held for people fleeing the Nazis. A lot of banks never returned the gold to the refugees or their families." His thoroughness put the people at the Sovereign Society conference to shame. "I met with a number of different ones, did some research, and went with a private bank that mostly deals with B people. But I'd recommend AIG Private Bank for you."

I asked him about asset protection, and he suggested a law

firm called Tarasov and Associates that most of his friends used. I trusted his recommendations more than the speakers at the conference, who seemed to view safety more as a marketing niche than a personal right.

I returned home emboldened by a new sense of mission and overwhelmed by the amount of work that lay ahead. I booked a trip to St. Kitts during Carnival in December, left a message for AIG to open a private banking account, e-mailed Grandpa for PT advice, and set up a meeting at Tarasov to hide my assets.

It was a good start.

# THE *GONE WITH THE WIND* GUIDE TO ASSET PROTECTION

If you wanted to withdraw your entire life savings and move it to a bank in Switzerland, what would you do?

Now that I'd decided to hide my assets offshore, the information from the Sovereign Society conference about the government tracking withdrawals and transfers of more than $10,000 applied to me. It seemed impossible to get the money from my American bank to the Swiss bank Spencer recommended without ringing alarm bells. Even if I moved it in small increments, there would still be a paper trail detailing exactly how much money I'd transferred.

So I did what any resourceful American would do: I bought a book on money laundering.

After all, it isn't a crime to move money secretly as long as the income's been reported to the IRS and any other necessary reporting requirements are met. And my intention wasn't to hide my earnings from the government, customs, or creditors, but to protect it from bank collapses, inflation, seizure, and lawsuits, which required leaving few traces of where it went.

Securing money overseas is not a new idea. Even in the novel *Gone With the Wind*, Rhett Butler keeps his earnings in offshore banks, enabling him to buy a house for Scarlett O'Hara after the

Civil War—in contrast to his Southern colleagues, who lose their fortunes due to blockades, inflation, and financial collapse.

For more practical, non-fictional inspiration, I bought Jeffrey Robinson's 1996 book *The Laundrymen*. I'd always wondered how empty video stores renting movies for $3 a day could stay in business, and why I'd see Russian thugs running clearly unprofitable frozen yogurt stands on deserted side streets. According to Robinson, it's because, in order to make illegal funds appear legitimate, crooks will slowly feed the money into the cash registers of a normal business.

"It's almost impossible to spot an extra $500 coming in daily through the tills of a storefront stocked with 15,000 videos," he writes. "Nor would anyone's suspicions necessarily be raised if that same owner ran a chain of twenty video rental stores and, backed up with the appropriate audits, awarded himself an annual bonus of $3.96 million."

Buried elsewhere in Robinson's book was the answer I was looking for. The best legal way to surreptitiously move money, it seems, is to buy something that doesn't lose its cash value when purchased. For example, there's a black market for people who transfer money by buying expensive jewelry, art, watches, and collectibles, then selling them in their destination country for a small loss—usually no greater than the percentage banks charge for exchanging currencies.

So once AIG Private Bank in Switzerland returned my phone call—assuming that, unlike Spencer's lawyer, they were actually willing to work with me—I planned to go shopping for rare coins.

But if it was all so legitimate, why did it feel so wrong?

While I waited to hear from the Swiss bank, I drove to Burbank to meet with the asset protection lawyers Spencer had recommended, Tarasov and Associates. The receptionist led me into a room with black-and-silver wallpaper where Alex Tarasov sat at a large mahogany desk with a yellow legal pad in front of him. With this pad, he would rearrange my business life forever.

"You did a very smart thing by coming here," Tarasov said. Twenty-five years ago, he had probably been a frat boy. Maybe even played varsity football. But a quarter century spent sitting at desks scrutinizing legal papers had removed all evidence of health from his skin and physique. "By taking everything you own out of your name, we can hide it from lawyers trying to do an asset search on you."

"So if they sue me and win, they won't be able to get anything?"

"We can make it very difficult for them to find the things you own and get at them. It's not impossible, but the deeper we bury your assets, the more money it's going to cost to find out where they are. And if we can make that time and cost greater than the worth of the assets, then you're in good shape."

Like Spencer had said, this was just insurance. The cost of setting this up would be like taking out a policy against lawsuits.

"So what do you own?" he asked.

I laid it all out for him. "I have a house I'm still paying for. I have some stocks and bonds my grandparents gave me when I was a kid. I have a checking and a savings account. And I have the copyrights to my books." I paused, trying to remember if I owned anything else. I thought there was more. "I guess that's about it. I have a secondhand Dodge Durango, I guess. And a 1972 Corvette that doesn't work."

In truth, I didn't own that much. But ever since my first college job, standing over a greasy grill making omelets and grilled cheese sandwiches, I had started putting money in the bank. Since then, I'd saved enough to live on for a year or two if I ever fell on hard times or just wanted to see the world. I didn't want

to lose the freedom that came from having a financial cushion and not being in debt for anything besides my house.

"Here's what we can do," Tarasov said. He then sketched this diagram on his legal pad:

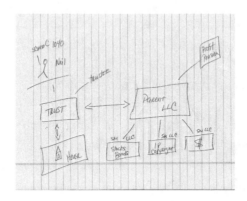

The stick figure was me. As for the boxes, I had no idea what those were. "These are boxes," Tarasov explained. I was clearly getting the asset-protection-for-dummies lecture. "Each box represents a different LLC"—limited liability company. "If we can wrap everything in an LLC, and then all those LLCs are owned by a holding company, and that holding company is owned by a trust that you don't even technically own, then you're safe."

I liked that last word. But I didn't understand the rest of it.

"So we're just basically making everything really complicated?" I asked.

"That's the idea. We'll even put your house in a separate LLC so that if someone trips and falls, they can't get at anything else you own."

When Tarasov was through explaining everything, I couldn't tell whether I was protecting myself from being scammed or actually being scammed myself. But I trusted Spencer, because he seemed too rich, too smart, and too paranoid to get taken in. So I told Tarasov to start wrapping me up in LLCs until my net worth was whatever spending money I had in my pocket.

"Once we have these entities set up, we can talk about transferring them to offshore corporations," Tarasov said as I left.

It sounded exciting, though I worried that by the time he was through charging me for all this, I wouldn't have any money left to hide.

Either way, my net worth would still be zero.

My next order of business was long overdue: to make my computer as secure as possible. That is, if my Internet searches for second citizenship options and private Swiss banks hadn't already caught the government's attention. For this, I enlisted the help of Grandpa.

He had responded instantly to the e-mail I sent after the Sovereign Society conference and, like a true money-minded PT, tried to sell me a series of books he wrote, *Bye Bye Big Brother,* for the bargain price of $750. Fortunately, I found a much cheaper abridged version his publisher was selling online.

In combination with a few Internet resources, I used the chapter titled "Secure Internet Communications" as my primer on computer safety. I spent an entire day downloading and installing encryption programs, firewalls, spyware destroyers, and software that supposedly enabled me to surf the Internet without being traced or tracked. Eventually, I hoped to use two computers: one solely for going online and a second laptop with all my data but no means of connecting to the Internet.

Though Grandpa's computer advice made me feel more secure, my long e-mail correspondence with him fanned other flames of paranoia. "Convicting someone [who's taking the steps you are] would bring glory to some little assistant D.A. or federal attorney," he warned. "The way out is not necessarily leaving at

once, but just be ready to move ass and assets at the first whiff of shit coming your way."

I thought back to a year earlier, when I felt like my life and livelihood were at the mercy of terrorists, the government, and the economy. Though those threats had grown even worse since then, I had become stronger, more stable, more informed, more resilient. All I needed to do now was visit Wendell in St. Kitts. Then I'd truly be able to leave when, as Grandpa put it, the first whiff of shit came my way.

I was almost free.

But, as Spencer continued to remind me every time we spoke, I still wasn't safe.

# WHERE TO FIND THE HAPPIEST ROOM ON EARTH

I have an office of one.

In that office is Tomas. I had hired him as an intern to help with the deluge of e-mails and obligations that came after *The Game* so I could be free to write other books.

In silence, Tomas watched me voraciously pursue my search for a second citizenship at the expense of almost all other work—until one day, as we ate lunch in the kitchen, he told me he had a wife.

"A wife? How's that?"

Not only was Tomas gay, but to make money, he placed ads on Craigslist offering, according to his headline, "Hairy Muscle Man Massages." The posting was accompanied by this picture:

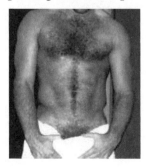

His fee was $100 per hour. Judging by the BMW convertible he drove, which far outclassed my Durango, business was clearly booming.

"I had to get married for my citizenship," he said. "She's crazy, but she's really good in the immigration interviews."

"So the whole time I've been working to get out of the country, you've been working to get in?"

"Yep." He smiled, exposing his sharp canine teeth and, along with them, a slight satisfaction at having successfully kept his secret from me.

Though Tomas was born in the Czech Republic, he was so Americanized that I'd never realized he wasn't actually a citizen. I guess assisting me a few hours a week was his way of having a legitimate job.

Five years ago, he said, he began the process. After talking to immigration lawyers, he decided that the quickest and easiest way to an American citizenship was through marriage. So he paid a female friend thirty-five thousand dollars to marry him. They even cosigned an apartment lease and opened a joint bank account to make it look official.

"Being American was always a fantasy of mine, ever since I was eleven," he said as he downed the dregs of a protein shake. I couldn't believe we'd never discussed it before. "I never felt like I belonged in the Czech Republic. I was always trying to escape this Eastern European subservient mentality."

As he spoke about it, I was overcome with a feeling of shame. There were millions of people living in conditions of extreme poverty all over the world, most of whom would sacrifice almost anything for a chance at a better life in America. I remembered the day the janitor of one of the buildings I grew up in became an American citizen. He told me he'd tried to escape from Romania five times by swimming across the Danube. Once, after he was caught, prison guards beat him and urinated on him for wanting to leave. As soon as he became a citizen here, he changed his name from Liveu Anei to the most American name he could think of: Lee Grant, after the Confederate and Union Civil War generals. Soon, he and his wife were investing in real estate, wearing matching jogging suits, and taking *Love Boat*–style cruises in an effort to be as American as possible.

"I love America," he'd always tell me when he was in a good mood, before his wife left him.

But what is America? In truth, it occupies less than one quarter of the geographical mass known as the Americas. Its proper name is the United States of America, though in truth it's hardly a name. France and Brazil—those sound like countries. The United States of America is more of an umbrella term, describing a hodgepodge of separate states bound together by an agreement, like a bundle of twigs tied with twine.

Even the flag, with its fifty stars and thirteen stripes, displays no single national identity, unlike the sole maple leaf of our neighbors to the north or the eagle eating a snake on the Mexican flag. The stripes represent the original thirteen colonies; the stars celebrate the fact that the country has since grown to fifty states. If there's a message here, it has to do with accumulation. In fact, no other national flag has that many stars, making us the best accumulators in the world.

Thus, perhaps my labors of the last year weren't anti-American but the most American thing I could possibly do. For my plan was not to forsake my American identity, but instead to start accumulating passports, citizenships, bank accounts, escape routes, and safe havens. I just wanted more stars on my own flag.

A few weeks later, just before I was scheduled to fly to St. Kitts, I accompanied Tomas to the convention center in downtown

Los Angeles to watch him get naturalized as a citizen. More than 4,000 people became Americans with him, followed by four thousand more in a ceremony immediately afterward. Most of the soon-to-be Americans were from Mexico, El Salvador, the Philippines, and, in a sea of black chadors, Iran.

Onstage, an Asian judge in a black robe and caftan sat at the center of a table, surrounded by uniformed immigration and police officers. An immense American flag waved behind them. The audience was asked to rise while several hundred soldiers, most of them Mexican-born U.S. residents, marched into the auditorium and filled the front rows. In exchange for their service to the country, they, too, would become citizens today.

Afterward, the judge told a story of how his father had moved from China to America for a better life. Then his captive audience, with photocopied handouts to help with the words, recited the pledge of allegiance and sang the national anthem. This was followed by one of the most touching scenes I'd ever seen.

"I hereby declare, on oath . . . ," the judge began.

"I hereby declare, on oath . . . ," repeated 4,000 people two minutes away from becoming U.S. citizens.

The judge continued, one voice leading 4,000: ". . . that I absolutely and entirely renounce and abjure all allegiance and fidelity to any foreign prince, potentate, state, or sovereignty of whom or which I have heretofore been a subject or citizen." Thousands of people sobbed as they spoke these words.

"That I will support and defend the Constitution and laws of the United States of America against all enemies, foreign and domestic." Hundreds more could hardly utter the words because of the immense, relieved smiles on their faces.

"That I will bear true faith and allegiance to the same." Others boomed the words passionately, as if their lives depended on them.

"That I will bear arms on behalf of the United States when required by the law; that I will perform noncombatant service in the armed forces of the United States when required by the law." Others seemed nervous, as if one wrong move would cause them to forfeit the rights they were about to receive.

"That I will perform work of national importance under civilian direction when required by the law; and that I take this obligation freely without any mental reservation or purpose of evasion. So help me God."

"You are now citizens of the United States," the judge announced. "Congrat—"

Before he completed the word, the room erupted with the sound of the largest exclamation point in downtown Los Angeles. Afterward, the new Americans looked around an auditorium full of strangers who had just become brothers. Tears involuntarily filled my eyes. For the Iranian woman in the burka standing next to Tomas, it had been fourteen years since she'd begun the process to become an American citizen.

It was as if 4,000 movies—full of hope, anxiety, ambition, tragedy—had just ended before my eyes. In that moment I realized that for every person who sees America as a villain, there's someone else who sees it as a hero. If anyone who wanted to decimate our country was in the room, they'd realize that it wasn't a president or a general or a businessman they were planning to destroy, but the dreams of hundreds of thousands of their own brothers, sisters, and countrymen.

Afterward, Tomas, in tears, filled out a passport application form, turned in his green card, and received his long-awaited naturalization certificate. Later that night, I joined Tomas and his friends for a celebratory dinner at a restaurant called Sur. The conversation soon turned from clubs and gossip to politics.

Tomas's housemate, an attorney, began talking about new

laws that, as he put it, "enraged" him—like the Military Commissions Act, which allows the government to declare U.S. citizens enemy combatants and imprison them without a trial, and the John Warner Defense Authorization Act, which gives the president the power to declare a public emergency and station troops in any city in America without permission from the local government.

"Get ready for martial law," Freddy, the attorney's much younger boyfriend, sighed.

"I read an article recently that said the voting machines in Ohio in 2004 were rigged," the architect continued.

"Yeah, and we go to other countries and preach about how they need to have free elections," Freddy interjected, "but we don't even have that privilege ourselves."

Suddenly, we all went quiet and looked at Tomas. It was his first day as an American citizen, and we were ruining it.

In that silence came my opportunity. I was finally able to ask what I'd been wanting to since he'd first told me about his quest. "In this climate, when America is doing so poorly economically and politically, especially since you already have a European passport, what made you so motivated to become an American citizen?"

Tomas didn't hesitate to respond. He'd probably known the answer long before he ever came to America. "It's not about freedom," he replied. "America is one of the least-free countries in the Western world. Things are so controlled here compared to Europe."

I had no idea what was coming next. Why would he want to become an American citizen if it wasn't for the freedom? Perhaps it was simply because his friends were here.

"I wanted to become a citizen for the opportunities," he finally continued. "In the Czech Republic, I had no future. In America, anything is possible. Anyone can become whatever he wants. It's all happening here. There are a million different paths and choices and careers open to everyone who lives in America. And no matter what happens politically, they can't take that away."

Everyone at the table fell silent. The truth has a way of doing that to people sometimes.

A few weeks later, as I sat in the airport, waiting for a flight to see my family on the way to St. Kitts, I thought about the words Tomas had spoken. One of the few pieces of advice my father ever gave me as a teenager was "Expect the best but prepare for the worst." Tomas expected the best. I was preparing for the worst.

LESSON 24

# HOW TO ALIENATE YOUR FAMILY IN SIXTY SECONDS OR LESS

For a brief moment in history, the lives of 80 million Americans hung by a thread. On the plane ride to Chicago to see my parents, I read Robert F. Kennedy's book about it, *Thirteen Days*.

The book describes the meetings that took place in President John F. Kennedy's inner circle during the Cuban missile crisis in 1962. For thirteen days, from the time the government learned the Russians were building a nuclear missile base some ninety miles off the U.S. coast in Cuba to the moment the Russians agreed to dismantle it, the world stood on the brink of nuclear war. According to historian Arthur Schlesinger Jr., in the event of an American invasion of Cuba—which many in the White House were advising—the Russians actually planned to fire the nukes at U.S. cities.

"One member of the Joint Chiefs of Staff, for example, argued that we could use nuclear weapons, on the basis that our adversaries would use theirs against us in an attack," Kennedy writes of proposed plans for a preemptive strike during those days. "I thought, as I listened, of the many times I had heard the military take positions which, if wrong, had the advantage that no one would be around at the end to know."

As I read the book, I wondered: If I had come of age during the Cold War, attended school assemblies where they showed films about what to do in case of a nuclear attack, been told that Russia had missiles pointed at my home city, watched helplessly while the House Committee on Un-American Activities ruined the lives of innocent Americans by accusing them of being Communists, and been at risk of involuntary service during the Vietnam War, would I have left the country?

I wasn't completely sure. So I asked my closest genetic match, my father. "Were you ever scared during the Cold War that something bad would happen to you or your family?"

"Not really," he replied. "I'd say that right now is the scariest time I've been alive."

Considering that my father had lived through World War II and the Cold War, in addition to serving as an army lieutenant in Korea, his answer gave some legitimacy to my concerns. "Why is it scarier now?" I asked him.

"In the Cold War, because of mutually assured destruction, you didn't think about the Russians sending their nuclear things over. But now, with terrorists, you don't know what's going to happen. They've killed innocent people everywhere on the planet. So it's scary at home and it's scary when you travel. Nothing's safe anymore."

I'd stopped in Chicago to have dinner with my parents on the way to St. Kitts not just to tell them about my escape plan, but to see if they wanted to be included.

"If you want," I offered, "I can look into adding family members to my citizenship application. This way, if anything happens in America, you have somewhere safe to go."

"We're fine," my mom said, dismissing the proposal as quickly as she'd dismissed most of my ideas, jobs, and girlfriends.

"So you don't want me to even check for you?" I asked, disappointed.

"Ask your father."

"I don't see the point," my dad said. "We're happy here."

"Are you sure? I mean, you just said you were scared." I wanted to give them another chance. Most people mistake comfort and familiarity for safety. Not until the flames are licking their rooftop will they leave—and even then they'll dawdle, trying to grab every last memento and stuff it into their car, as if their possessions held the very essence of their identity. "I'll do the work. All you'd need to do is sign the papers."

"Just leave us out of it," my mom said. The conversation was over.

"Okay, well, if you're ever in trouble, you can come stay with me anyway."

I'd wondered at times why I seemed to be the only one of my friends pursuing safety so voraciously. As I sat with my parents, I realized that it was in part because of my upbringing. My parents were worst-case thinkers, and used to constantly warn me about everything that could go wrong when I grew up and left home.

They also practiced what they preached. When we went on vacation, my brother and I weren't allowed to tell our friends, so that no one knew our house was empty. And we'd leave for the airport in shifts: my father and I would take a taxi as my mother and brother pretended to say goodbye, and then they'd follow in another cab shortly afterward. While we were gone, the house lights were hooked up to timers so they'd go on and off as if we were still there. In short, I was raised in a world where strangers were enemies and privacy meant protection.

"You know what you might find interesting?" my father said, playing peacemaker. "There's a book that came out when everyone was scared of the Russians. Someone bought it for me as a joke. It tells you how to build nuclear shelters and radiation suits."

"I'm not that paranoid," I told him, then added, "Do you still have it?"

"I'm pretty sure I threw it away. But it was called *Life on Doomsday* or something like that."

After dinner, I went online and found the book, which was actually called *Life after Doomsday*. The author, Bruce Clayton, had written it during the tail end of the Cold War, and it quickly became somewhat of a bible for the burgeoning survivalist movement. He'd also written a sequel in response to 9/11, *Life after Terrorism,* which I ordered as well. Spencer's obsession with guns, planes, and his latest and most outlandish escape plan, submarines, kept echoing through my head.

Later that night, I asked my brother if he wanted to get a second passport with me. At first he declined, saying he was worried there'd be negative consequences. After I reassured him, he said he didn't have the money for the citizenship. So I told him I'd take out a loan and pay it off for him. Then he flatly declined.

I was alone.

There is a phenomenon known as social proof. It's the idea that if everyone is doing something—whether it's buying a certain pair of shoes, seeing a particular movie, or criticizing some unfortunate individual—it must be right. So if no one else was trying to escape the country besides me, then it must be wrong.

The problem with that logic is that by the time everyone is doing it—after the next terrorist attack or economic depression or political clampdown or epidemic disease—it will be too late. And those who want to escape will be left trying to barter money, sex, connections, and everything else they have just for the privilege of living.

# HOW TO SURVIVE COLD WEATHER

DIG A TUNNEL INTO A SNOW BANK, LEADING TO A COMPARTMENT LARGE ENOUGH TO SIT UPRIGHT IN. PAD THE GROUND WITH BRANCHES AND LEAVES, IF AVAILABLE. BLOCK THE ENTRANCE WITH YOUR BACKPACK OR SNOW. ALTERNATELY, IF YOU FIND A TREE WITH LOW-HANGING LEAFY BRANCHES, DIG A PIT AROUND THE BASE OF THE TREE, AND COVER THE TOP AND BOTTOM WITH FOLIAGE. REMEMBER THAT WIND CAN KILL YOU BUT SNOW WALLS WILL INSULATE YOU.

NINE MONTHS EARLIER...

# HOW TO DEFEAT SECURITY CAMERAS AND BARBED WIRE

TO EVADE CAMERAS, EITHER SHINE A LIGHT OF 130 LUMENS OR MORE DIRECTLY ONTO THE LENS OR JUST SHOOT THE LENS WITH A PAINTBALL GUN. FOR BARBED WIRE, PUT AN OPEN CARDBOARD BOX OR FLOOR CAR MAT OVER THE TOP OF THE FENCE, THEN CLIMB OVER THAT AREA OF THE FENCE. IF YOU CHOOSE TO CUT THROUGH THE FENCE, REPLACE THE BROKEN WIRE WITH KNOTTED PARACORD SO IT LOOKS INTACT FROM A DISTANCE.

SIX MONTHS EARLIER...

# HOW TO ESCAPE A MOB

LEAVE A CS GRENADE IN A ROOM OR ALLEY BEHIND YOU OR THROW IT IN THE ENEMY'S CAR TO CREATE A NON-LETHAL CLOUD OF GAS THAT WILL INCAPACITATE PURSUERS.

THREE MONTHS EARLIER...

# HOW TO CREATE FIRE FROM A BATTERY

TO START A FIRE FROM A CAR BATTERY, ATTACH YOUR JUMPER CABLE TO THE TERMINALS. TOUCH THE ALLIGATOR CLIPS ON THE OTHER END TOGETHER JUST OVER YOUR TINDER BUNDLE TO CREATE A SPARK THAT WILL IGNITE IT. TO START A FIRE FROM A NINE VOLT BATTERY, RUB THE TERMINAL ENDS WITH STEEL WOOL UNTIL IT BURSTS INTO FLAMES. COTTON BALLS DIPPED IN VASELINE (OR CHAPSTICK) MAKE FOR EXCELLENT TINDER.

TODAY...

# HOW TO NAVIGATE WITHOUT A COMPASS

PLACE A THREE-FOOT-TALL STICK VERTICALLY ON A FLAT STRETCH OF GROUND. MARK THE END POINT OF THE RESULTING SHADOW WITH A ROCK. WAIT TWENTY MINUTES, THEN MARK THE NEW END POINT OF THE SHADOW WITH ANOTHER ROCK. A STRAIGHT LINE DRAWN BETWEEN THESE TWO POINTS WILL RUN ROUGHLY WEST TO EAST, WITH THE FIRST POINT BEING WEST.

IF YOU HAVE A WATCH, YOU CAN ALSO NAVIGATE BY HOLDING YOUR WATCH HORIZONTALLY, WITH THE HOUR HAND POINTING AT THE SUN. THE MIDPOINT BETWEEN THE HOUR HAND AND THE NUMERAL 12 ON THE WATCH WILL BE SOUTH (UNLESS YOU'RE IN THE SOUTHERN HEMISPHERE, IN WHICH CASE IT WILL BE NORTH).

# HOW TO MAKE A BOW AND ARROW

CUT A FOUR-FOOT LENGTH OF PVC PIPE AND CUT SLITS TWO INCHES FROM BOTH ENDS OF THE PIPE. EACH NOTCH SHOULD GO HALFWAY THROUGH THE PIPE. CUT A LENGTH OF STRING ROUGHLY FOUR INCHES SHORTER THAN THE BOW. TIE A SMALL LOOP IN BOTH ENDS, AND SLIP A LOOP INTO ONE OF THE NOTCHES. BEND THE BOW, AND LOOP THE OTHER END OF THE STRING INTO THE OTHER SLIT. IF POSSIBLE, USE ELECTRICAL TAPE TO MAKE A GRIP. SHARPEN ONE END OF A STRAIGHT STICK (OR WOODEN DOWEL) AND CUT A NOTCH IN THE OTHER END TO MAKE A SIMPLE ARROW.

TO BE CONTINUED...

## LESSON 25

# PROOF THAT A FOOL AND HIS MONEY ARE SOON PARTED

I t was here, in the streets of Basseterre, where the debris of carnival rotted in the sun and teenage boys leaned against the same peeled-paint walls their fathers once did, that I would make my stand.

In the center of Market Street sat the office of Maxwell Webb and Partners. My entire plan rested solely on his shoulders—and, until just two hours earlier, I'd never even heard his name.

Maxwell greeted me at the door, alone in the office. He shouldn't even have been there during carnival. He made sure to remind me of that. Several times.

I wanted to trust him, because I had to. But the diamond-encrusted ring on his middle finger, emblazoned with the initials MW, told me not to. It told me that he wanted to be respected, to be considered a player, to chase status and women. And in a poor town of fifteen thousand people, that wasn't difficult. A ring like that was probably all it took.

And I was about to contribute $2,500 toward his next ring.

He sat down heavily in front of a large walnut desk covered with thick file folders. I tried to memorize the names on them—Bernard Hellick, Thomas Murgic—so I could do some independent research and see what kind of clients he dealt with.

163

He stared quietly at me, his egg-shaped face bearing the first signs of growing jowls, sweat stains materializing in his white button-down shirt. He knew why I was there. But it appeared I must tell him anyway.

"So how does the process work?" I asked.

When I'd first arrived in St. Kitts, Wendell had told me what needed to be done. But he'd sent me here, to this office, to this stranger with the expensive ring, to get it done. And I needed reassurance. I had no idea if this move was smart and prescient, ultimately saving my life and hopefully that of my family—or if it was the stupidest thing I'd ever done, a testament to paranoia that would end in my own bankruptcy.

"You'll need some property," Maxwell said. "You can call Victor Doche at St. Christopher Club, Ron Fish at Half Moon Bay, or Nicholas Brisbane at Calypso Bay."

The office fan spun uselessly above his head, circulating heat, as he read their numbers to me from the local telephone directory. I asked about other options. I wanted to think carefully about this. But he told me I had no other options. He was right.

"Once you pay the twenty-five-hundred-dollar retainer fee, I'll give you the papers you need to fill out."

I didn't like him. I wanted someone to hold my hand, to explain things rather than just tell me what to do. I like to understand something before committing to it. I don't think that's wrong.

But either Maxwell wasn't the talkative type, or he saw that I was by no means the kind of person he'd want to align with. His walls were filled with framed letters, photographs, and certificates testifying to the high opinion he wanted others to have of him. But I didn't smell like money and power and a thick file folder and a photograph to put boastfully on the wall. I smelled like the inside of an airplane, economy class, middle seat.

"How many of these cases do you take a year?" I asked him, looking for some reason—any reason—to have faith in him.

"Ten," he said.

"And how many applications does the government take each year?"

"About a hundred. Ninety-nine percent of them get approved."

So Maxwell handled 10 percent of the traffic. That was reason enough for me.

I wrote him a check for $2,500. He handed me a sheaf of papers. One was an application for citizenship. The other was an application for a passport. In six months, Maxwell told me, if all went well, I would officially be a citizen of the Federation of St. Kitts and Nevis in the Caribbean Commonwealth.

Spencer's words burned in my ears. I was buying insurance. Insurance against America, the country I was born in, the country my friends and family live in, the country I love.

This wasn't supposed to happen in my lifetime.

# THE ART OF DISAPPEARING

The next day, I woke up early, found a taxi, and went house shopping.

The driver's name was Chiefy. He was a short, jolly Kittitian with a smile designed to win the trust of tourists.

"Most of the land in this spot was owned by a Russian gangster who couldn't get citizenship anywhere in Europe," he informed me as we swerved around a large green hill. Every time he spoke, he turned around to make eye contact, which would have been polite if it weren't for the narrow lanes and tight turns of the coastal road.

"Except that hill right there," he went on. "That's owned by a Saudi Arabian criminal who stole his son from his ex-wife and brought him to America. He bought the property in exchange for a diplomatic passport, so he could safely get to the States and back to see his son."

And so began a tour of the checkered past, promising future, and eccentric characters of the island that might one day be my home. Discovered by Christopher Columbus in 1493, St. Kitts was united with a neighboring island, Nevis, four centuries later to become the smallest country in the Americas and one of the best-kept secrets in the Caribbean.

Chiefy's first stop was the Rawlins Plantation, where he introduced me to Kevin, the owner, a short Welsh man with buck teeth and shifty eyes. From there, Chiefy and I moved on to Calypso Bay, Half Moon Bay, a row of what looked like concrete railway containers in the infancy of their construction, and half a dozen other places—each owned by a character more peculiar than the last, most of them small-time operators with big dreams of striking it rich in real estate on the developing island. It was the Wild West, but with an ocean view.

The last place I looked at was St. Christopher Club, a small enclave of three-story white buildings on a narrow strip of land separating the Atlantic and Caribbean beaches.

"You know, of the last twenty-five units, sixteen sold to Americans and every one of them was in the citizenship program," the co-owner, Victor Doche, a short, garrulous man with a Charlie Brown smile, said as he walked me through a newly built apartment.

"Is that normal?"

"No, it's new."

"Who bought them before?"

"Originally, it was because of Idi Amin." I wasn't sure what this meant—though I later learned he was referring to Amin's expulsion of the entire Asian population of Uganda following a dream in which God told him to banish them. "Then it was Russians. A lot were very powerful people. Not all of them were in the Mafia, though. I bought my apartment back from a guy who was a general in the KGB.

"Now it's all Americans," he continued. "One family is here because they fear war with Iran. Another because they don't want their children drafted. And another is here because they make money offshore and don't want to bring it back through the U.S. and have it taxed."

"Interesting." So I wasn't alone. I was part of a movement, an awakening, a disillusionment. The great American leaking pot.

We were running away from war and death and taxes. If our instincts were right, we'd be rewarded with life. If they were wrong, we'd be stuck on one of the most beautiful islands I'd seen. There was no downside. Or so I thought at the time.

Victor was a similar seeker. He was born in Egypt, but, he explained, "as a Christian, I saw that things were getting bad for my people in the seventies and left." He fled to Montreal, where he helped bring aerobics to the country with help from Jane Fonda. As he told his story, I was reminded of my silent pledge in Mrs. Kaufman's class, and I realized that running doesn't always mean hiding. Sometimes it means growing.

"The St. Kitts and Nevis passport is very good," he continued. "Because the islands are part of the British Commonwealth, people use them as a backdoor entrance to Great Britain. I'm told that if you lose the passport in a foreign country and go to the British embassy, they'll replace it with a British passport."

So not only was I going to become a Kittitian, I was going to become a member of the British Commonwealth. Maybe I could get my coveted EU passport after all.

After looking at one of the apartments, a tranquil duplex with views of both coasts, I walked to the Atlantic beach a few yards away. The weather was warm, but not hot, with a mild breeze. It felt as if this were the temperature at which the human body was designed to live. The waves crashed white and warm against the shore. There wasn't another person in sight. It was the antithesis of my life in cities up to that point.

I recalled the five criteria Spencer and I had selected as requirements for a second citizenship: credible passport, stable government, minimal tax liability, maximum two-year wait for a citizenship, and warm climate with beaches.

St. Kitts met every criterion.

I crossed the property, walked two hundred yards to the Ca-

ribbean coast, and ordered a rum punch at an open-air bar on the beach. The bar's co-owner, it turned out, also lived at St. Christopher Club.

His name was Regan, and he told me that the prime minister of the island had a spare apartment in the complex. He also said there were plans to build a Ritz-Carlton next door, which would be good for property values.

A new life seemed to await me here, free of petty concerns and complications. The slow, laid-back pace alone would probably prolong my life. And if I ever got caught in the system, all I had to do was knock on the prime minister's door with a bottle of wine and ask for advice. I couldn't imagine a better oasis for a technological nomad.

# LESSON 27

# WAR, GENOCIDE, AND OTHER REAL-ESTATE SCAMS

When I took a cab to the island's capital, Basseterre, at five A.M. the day after Christmas, the J'ouvert street party—the climax of the island's two-week-long carnival celebration—was in full swing. Bands lined the sides of floats and trucks, hammering away on steel drums as hundreds of drunk revelers danced ecstatically behind them. It was not a show for tourists, but an age-old rite occurring despite my presence. It seemed hypocritical to call this place my home when these were not my people and this was not in my blood.

I didn't belong here. Kevin didn't belong here. Victor didn't belong here. Regan didn't belong here. The Russians, Ugandans, and all the eccentric real-estate developers with their big dreams didn't belong here.

Then again, there was a time when these revelers didn't belong here either. The Kalinago tribe fought the Igneri tribe for the island. The French and the Spanish fought first the Kalinago, then each other. The French and the British then fought, uniting briefly to massacre nearly the entire native population at a site now known as Bloody Point.

Yet still the island belongs to no one. It's just a neutral witness to human nature. And all that blood—like most of the blood spilled over man's inability to cohabitate and share—has been shed for nothing. So until the battle resumes one day, the island's guests will spend their time exchanging pieces of paper that represent value for other pieces of paper that represent land, just as we do everywhere else in the world. More than warfare, this symbolic paper, a single sheet of which can make a man a slave or a king, is the pinnacle of human civilization.

And that afternoon, I had my final meeting with a member of the race of man dedicated to the mastery of that paper: my lawyer.

"I can deal with bastards, but I hate assholes," Maxwell was telling someone on the phone when I entered. "Some people, they shouldn't be dealt with nicely."

When he hung up, I tried to make small talk and asked how his holiday had been so far. This was a mistake.

"I've been working the whole time. I can't take a vacation like you." He seemed to feel he was the only person in the world who had to work. "It's because of this thing." He pointed to the mobile phone hanging from his belt. "It never stops."

I pulled the forms he'd given me out of an envelope I was carrying. I'd filled in my name, address, and my parents' address. But other questions—asking for my occupation, annual income, and reason for wanting the citizenship—I'd left blank, because I had no idea what the government was looking for. If it wanted to attract rich investors to the island, it would probably reject me because I was just a writer and my income was nothing in comparison with the business magnates and Russian mobsters who usually applied for citizenship. So I asked Maxwell for advice.

"Is it better to put writer or author?"

"Doesn't matter."

"How should I describe my source of income?"

"They just need a guarantee that you have an income."

"Okay. One last question, then I'm out of your hair: what should I put down where it asks 'reason seeking citizenship'?"

He groaned as if he'd rather be in Guantánamo getting tortured and uttered with great effort, "'Alternative citizenship and future retirement home.'"

As I filled out the rest of the application, he yelled something about making a mistake on my invoice at a beleaguered secretary, who was wearing inordinately high red peep-toe shoes. She returned a few moments later with the largest bill I'd ever been handed in my life:

## Maxwell Webb & Partners
### Barristers-at-Law - Solicitors
### Business and Property Law - Public Notary

CLIENT: NEIL STRAUSS

Date: 26th December

| ITEM | DISBURSEMENTS | | FEES | | TOTAL | |
|---|---|---|---|---|---|---|
| Assurance Fund | $978 | 00 | | | $978 | 00 |
| Registration Fees | $2 | 68 | | | $2 | 68 |
| Legal Fees on Transfer | | | $13,725 | 00 | $13,725 | 00 |
| Fees for pre-approval | $2,500 | 00 | | | $2,500 | 00 |
| Citizenship Fees | $35,000 | 00 | | | 35,000 | 00 |
| Passport Fees | $32 | 00 | | | $32 | 00 |
| Certificate of Registration Fee | $47 | 00 | | | $47 | 00 |
| Legal Fees on Citizenship | | | 15,000 | 00 | $15,000 | 00 |
| | | | | | | |
| TOTAL AMOUNT OF BILL | | | | | $67,284 | 68 |
| AMOUNT CREDITED-RETAINER | | | | | $2,500 | 00 |
| TOTAL NOW DUE | | | | | $64,784 | 68 |

I didn't have that kind of money lying around.

"I'll need you to wire that to my account," Maxwell informed me, "along with the full cost of the apartment."

"I thought I was just supposed to pay a ten percent deposit." Cold sweat began prickling my forehead and the back of my neck.

"The government likes to see all the money in an account so they know you won't back out."

Until that moment, I hadn't given much thought to the practicalities of affording the citizenship. Perhaps if I took out a second mortgage on my current home, I might be able to raise that

amount. Then I could take care of the monthly mortgage payments by renting out the unit when I wasn't on the island. So not only would I become a citizen of St. Kitts, I'd also become a foreign real-estate speculator and shady landlord.

But it was still a devastating amount of money. What if I was a victim in some sort of long con? Maxwell had the home-team advantage. I was completely at the mercy of his word.

He irritably shuffled through my documents and handed me a piece of paper that would, hopefully, become my passport one day. "Write your name inside the box, but make sure it doesn't go outside the box," he told me, then repeated it, as if I were stupid.

"In order to submit your application," he continued, "I'll also need a negative HIV test, a clean police record, nine passport photos, and a copy of your birth certificate."

I handed him the lease agreement for St. Christopher Club and asked him if it was okay to sign.

"I did all the documentation for St. Christopher Club," he replied, as if offended I'd dare to ask such a question.

"So you wrote this agreement, then?"

A vertical tremor in his face seemed to imply his assent. This was quite possibly the stupidest thing I'd done in my life.

"So," I asked, hoping for more reassurance than at my last visit, "what's the likelihood of this going through?"

"As long as you have no criminal record, a clean HIV test, and no tax problems in the U.S., you're fine," he replied.

"Anything else ever go wrong?"

"As long as you have no criminal record, a clean HIV test, and no tax problems in the U.S., you're fine."

It was clear he just wanted me out of his office. Foolishly, I tried to make more small talk in an effort to befriend him. "Hope you get some free time in."

"I have another client coming," he responded. "Can I ask you to leave?"

I walked out with a sinking feeling in my chest.

In the Andrei Tarkovsky film *Stalker*, just before making a fateful decision, one character says to another, "There must be a principle: never do anything that can't be undone."

That principle is why, on the brink of a big decision, the first thing to fill my mind is doubt. What separates the strong from the weak, I reminded myself as I wandered sticker-shocked through the streets of Basseterre, is the ability to act instead of spending most of life paralyzed, too scared to make a choice that might be wrong.

St. Slim Jim, patron saint of reckless decisions.

## LESSON 28

# CALCULATE THE ODDS THAT YOU'RE IN JAIL RIGHT NOW

That night, weakness struck. In a few days, I'd be committed to an expense of over half a million dollars, which was more money than I had.

And what was it all for? Symbolic paper. A passport, which is just a teeny little booklet that means nothing to the universe. Realistically, the world wasn't likely to end in my lifetime. And if it did, everyone on St. Kitts would be just as dead as everyone in America.

If there were a smaller-scale world disaster, things would probably be even worse on an island in the Caribbean, where I was more likely to be a victim of food shortages, droughts, hurricanes, blackouts, and tsunamis. There's nowhere to run and nowhere to hide on an island—especially one in the smallest country in the Americas. I'd become so focused on my search for a passport—so consumed with escaping the blowback of American politics—that I'd forgotten the survivalist lessons I'd learned on Y2K and 9/11.

Soon, the whole endeavor began to seem like the biggest travesty ever. If something horrible happened in America, would a St. Kitts passport even get me out during a state of emergency? What if it was confiscated by customs agents? Or what if Victor,

Maxwell, and Wendell were in collusion and just ripping me off? I didn't have anyone to protect me here.

Once I'd ridden out that wave of anxiety, a new one formed. I began worrying that I'd blabbed my name and occupation to too many people. If they Googled me and saw the filth I'd written, they might not sell me the apartment or give me a citizenship. And then I'd be stuck in America if anything bad happened.

And so it went, all night, one wave of anxiety after another—half of them spent worrying that I wouldn't get a passport, the other half spent worrying that I would.

I fell asleep around dawn for a few fitful hours, until I was woken by my cell phone. AIG Private Bank was finally returning my call.

Every day, my small savings were dwindling as the dollar dropped relative not just to the euro, but even to the Caribbean currency here. I never thought I'd see the day when Eastern Europeans came to the United States for the cheap shopping.

"I'd like to inquire about opening a private banking account," I told the woman.

"Great," she said, with barely a trace of a Swiss accent. "Let me ask you a few questions."

"Sure."

"Are you an American citizen?"

"Yes, I am."

"We don't deal with American citizens for a few years now."

"But my friend Spencer Booth is American, and I think he has an account with you."

"It's likely an older account. We don't do business with American citizens anymore. Sorry, good-bye."

Before I could respond, she had hung up. I felt like an outcast. I couldn't believe a bank wouldn't take my money solely because I was American.

I'd noticed that many of the banks I'd researched had special policies for dealing with United States citizens. Even some of the online companies selling vintage travel documents said they no longer shipped to America because U.S. customs agents were opening and confiscating the packages. The government seemed to be sticking its nose everywhere.

In the meantime, I'd discovered a few other interesting facts: According to a report issued by Reporters Without Borders, the United States was ranked as having the fifty-third freest press in the world, tied with Botswana and Croatia. According to the World Health Organization, the United States had the fifty-fourth fairest health care system in the world, with lack of medical coverage leading to an estimated 18,000 unnecessary deaths a year. And according to the Justice Department, one in every thirty-two Americans was in jail, on probation, or on parole.

Rather than having actual freedom, it seemed that, like animals in a habitat in the zoo, we had only the illusion of freedom. As long as we didn't try to leave the cage, we'd never know we weren't actually free.

That phone call was all it took to let me know I was doing the right thing.

Before going home, I had dinner with Wendell at a restaurant called Fisherman's Wharf and thanked him for his help.

After the meal, he patted my shoulder and smiled. "Next time I see you, you'll be a citizen of St. Kitts and Nevis just like me," he said. "When you get married, your wife will be a citizen. And when you have kids, so will they."

He stepped into his SUV, started the engine, then unrolled the window and concluded his thought: "One day," he said, beaming, "when you come back to America, no one will recognize you. You'll be a Kittitian."

At the St. Kitts airport the next morning, I felt like I was returning not to a country but a fortress. "Your country is so tough to get into," the ticket agent complained as she checked my documents for the flight home. "They make it so hard for us."

She looked up at me and said it louder, almost with venom, as if it were my fault. *"They make it so hard for us."*

She wasn't alone in her opinion. A survey released the previous month by the Discover America Partnership had found that international travelers considered America the least-friendly country to visit.

"That's why," I told her, with the newfound pride that Wendell had instilled in me, "I'm moving here."

LESSON 29

# WHY THE GOVERNMENT FROWNS ON SMILING

I flew back to Los Angeles, determined to gather the material I needed for my citizenship application as quickly as possible.

To obtain my criminal record, I sent a letter to the FBI, along with a set of rolled-ink fingerprints and a certified check for eighteen dollars. The government claimed that it wouldn't keep my prints on file after the process was complete, and I hoped this was true. Though I didn't plan on committing any crimes, there may come a time when I want to escape with my second passport, and I wouldn't want my fingerprints to give my identity away.

Perhaps I was more like the paranoiacs at the Sovereign Society conference than I cared to admit.

To get my passport pictures, I visited a small photography services store in Koreatown. "Don't smile," the owner instructed me. "The government doesn't like smiling in them these days."

Between the black shirt I was wearing and the grim expression on my face, I looked like just the kind of mobster who would create a new identity in St. Kitts.

My next errand was the HIV test. I found it strange that AIDS was the only disease the Kittitian government worried about. And though I was always safe, the problem with health exams is that they open possibilities one doesn't consider before taking them. After all, every test has the possibility of being failed. If I were HIV-positive, then this whole attempt to save myself would be pointless. In the grand scheme of things, health trumps nationality every time.

Fortunately, my white blood cells were as clean as my criminal record.

My last task was the one I dreaded most: paying the bill. I called the mortgage agent who'd helped me get my home in L.A. She sent an appraiser to my house, estimated the value, and loaned me the money I needed, no questions asked. This was before the mortgage market crashed and she was forced to fire her staff, sell her dream home, and watch helplessly as her annual income dropped from seven figures to five.

Rather than being exciting, the loan was terrifying. If my quest was for freedom, going deeper into debt was the worst way to attain it.

Finally, I put everything in an envelope and sent it to Maxwell, hoping he wouldn't abscond with the money and go golfing somewhere. My future lay in his hands.

As I waited for him to submit the application, I continued my search for a Swiss bank account. After getting shut down by AIG, I'd contacted half a dozen other banks in Switzerland. Every one of them had also turned me down. So I decided to dig deeper and figure out why no one would take my money.

I began calling every Swiss bank I could, until I finally found one that accepted clients from the United States. The name of the institution: Arab Bank. Unfortunately, it didn't accept initial deposits of less than half a million dollars.

Frustrated, I pulled out my notes from the Sovereign Society conference and called the Swiss branch of Jyske Bank.

"I'd like to open an account for you, but if we take a U.S. citizen, we risk your government closing our banks in the United States," a male employee named Kim informed me. "But we have clients from one hundred and eighty countries, so at least we can still serve one hundred and seventy-nine of them."

"Why won't anyone do business with Americans?"

A few years ago, Kim explained, the United States started requiring Swiss banks to designate the American government a "qualified intermediary"—which means that the bank has to report information on American clients to the U.S. government, withhold a percentage of interest paid to the account, and file tax forms with the IRS. In addition, the U.S. now required Swiss banks to register with the Securities Exchange Commission if they wanted to do business with American citizens.

Jyske chose not to register, Kim said, because "it will compromise secrecy."

I couldn't believe what I was hearing. Bank privacy for a U.S. citizen, like many other freedoms, was basically a thing of the past.

"It's a strange time to be an American," I told him.

"It definitely is." I felt like he wasn't just saying that to be polite, but he actually sympathized.

Spencer had once given me a list of signs that meant it was time to leave a country. They included the government sealing its borders, banning the press, or forbidding citizens to move money offshore. The Nazi regime, for example, made it illegal to have foreign accounts. One of the reasons the Swiss originally drafted bank secrecy laws was that three Germans were executed after the Nazis discovered they held foreign bank accounts.

"Can you do business with people from St. Kitts?" I asked Kim before hanging up.

"Of course."

"Good to know."

As the weeks passed and I waited to hear from Maxwell, I wondered what else I needed to do to prepare for a future visible only through gathering storm clouds.

Guns. Planes. Submarines.

Spencer's words kept coming back to me. And though I wasn't about to take a third mortgage on my house to buy a submarine, perhaps I was avoiding the real work. Because it meant reinventing not just my nationality, but reinventing myself.

Guns. Planes. Submarines.

It meant not just becoming like those I made fun of, but becoming even more extreme than them.

Guns. Planes. Submarines.

I wasn't that crazy. I'd found my island paradise. That was all I needed. Or so I thought—until the smallest disaster in the world changed my mind.

## LESSON 30

# HOMEWORK: FIND A STRANGER YOU TRUST

*St. Christopher Club, St. Kitts, 2007*

*I am writing this on the dying battery of my computer. It's been several months since I first visited this island and, though I have a place to live here now, I still don't have my citizenship.*

*There was another blackout tonight. There are no backup generators, and the emergency lights only last fifteen or so minutes.*

*All the food in my refrigerator has spoiled again, and not only do I not have any matches for the candles, but I don't know where to get any.*

*It's a holiday today, and everything nearby is closed. I don't have a car, because the rental companies don't have any left. I know this because I called every single one of them yesterday.*

*I hear footsteps in the hallway and voices outside. If someone came up here, it wouldn't be hard to break in. There's just a flimsy lock, like on the doorknob of a bathroom.*

*I don't know if I have any neighbors. I haven't seen a person in the complex the whole time I've been here, outside of the local workers, who always seem to be around.*

*The apartment is on the flat part of the island, a little stretch of a square mile or two called Frigate Bay, home to most of the resorts and casinos and tourists. On the hill overlooking us is Basseterre, where the locals live in small town-houses and shanties. There are gangs there, and crime and murder.*

*The snap can occur here too, just as it can in the United States, just as it has in the past in Haiti and China and the Ivory Coast. One little snap—be it due to hunger, unemployment, propaganda, or simple resentment—and they'd all come running down here with the machetes they once used to harvest sugarcane. It would be over in a matter of minutes.*

*The police wouldn't help us. They, too, live up on the hill. And their loyalty is to the hill, not to the rich foreigners who treat them like servants.*

*The American government wouldn't help us, because it is far away and I am a traitor.*

*There is only one solution if I am to stay here: I will have to stock up.*

*Spencer was right. I've become so obsessed with repatriating that I haven't been looking at the big picture. A passport will help me escape, but it won't help me survive. I can't eat it, drink it, or defend myself with it. I can only run with it.*

*To actually survive on my own, I'll need emergency supplies. Plenty of them. Because if the power keeps going out on a regular day, I can't even imagine what would happen during an actual disaster.*

*I'll need guns too. To protect myself. From marauders who want my emergency supplies and my surfboard and my PlayStation.*

*And, more than anything, I'll need skills. I'll need to learn to live on my own without electricity, running water, gas. I'll need to learn to truly be sovereign.*

*The lesson of Katrina wasn't that the United States can't protect its own. It was that no country can protect its own.*

*No place is safe and no government can guarantee the well-being of its citizens.*

*There's only one place to find true safety: from within.*

## PART FOUR

# SURVIVE

I must find the home of the Faraway.
He is a human being just as I am.
Yet he has found everlasting life . . .
Surely he can teach me how to live for days without end.

—*Gilgamesh,* Tablet IX, 2100 B.C.

# SAFETY GUIDELINES FOR THE USE OF HORSES, GUITARS, AND TOWELS

I like the way you're thinking," Spencer said. I was telling him about my epiphany during the blackout in St. Kitts. "You have to make a thorough plan, though."

"What do you mean?"

"Do you have guns yet?"

"No."

"Can you fly a plane?"

"No."

"Well, if the system breaks down, you won't only have to worry about surviving in St. Kitts. You also have to worry about how you're going to get there." I could hear the ring of satisfaction in his voice. He could tell my attitude had changed. "There won't be major airline flights or cell phone reception or probably even working gas pumps."

"Good point. I'll try to get out ahead of time then."

"But what if there's no warning?"

The so-called system is something we take for granted. We depend on it to give us an inexhaustible supply of electricity, water, food, gas, Internet, phone service, garbage removal, long-distance transportation, civil order, twenty-four-hour convenience stores, and *Seinfeld* reruns. But what would happen if it stopped working—and, suddenly, there was nothing to depend on?

"My feeling is that instead of evolving constantly toward a more advanced civilization, human history is cyclical," Spencer was saying. He had me in the palm of his hand now. His anxieties were my anxieties. "And, just like Rome and Egypt and other advanced civilizations before us, we're past our zenith. We're growing weak, while the tribes that want to destroy modernity are growing stronger and more committed."

To Fliesians, civilizations don't keep evolving. They progress until some reactionary element hits the reset button, and they have to start all over again.

"Someday," he concluded, "a future civilization is going to find our computers and hard drives and have no idea they contain the entire history of our society. They'll just think they're funny-looking rocks and use them as tools."

After talking to Spencer, I walked to my bookshelf and pulled out the two Bruce Clayton books I'd ordered after dinner with my parents—*Life after Doomsday* and *Life after Terrorism*.

As I leafed through them, I began to feel stupid for having thought that being independent of America meant just having a nationality and a bank account somewhere else. I needed to become independent of everything. All my life, I had thought that freedom was something that, as Americans, we were privileged to have, thanks to the Constitution and the Bill of Rights. But those documents didn't create freedom. They created a system. And systems create dependencies. Real freedom, I realized, meant knowing not just where to go, but how to take care of myself if the system ever broke down.

And since returning from St. Kitts, I didn't have just myself to look after anymore. I'd started dating someone. Her name was

Katie. She was a quick-witted, high-spirited Russian–American heartbreaker with two-tone black-and-blond hair and a penchant for midriff tops, tight blue jeans, and push-up bras. One night I saw her sing Leonard Cohen's "Famous Blue Raincoat" while her sister, Grace, accompanied her on guitar. From that day forward, I was smitten.

In my survivalist fantasies, I'd imagined dating a rugged GI Jane type of girl who'd grown up on a farm and would teach me how to milk cows and raise chickens. Instead, I'd found someone with even less outdoor experience and more irrational fears than myself.

Katie was scared not just of driving a car and flying in a plane, but of literally everything—from guitars (she was afraid a string would break and lash her in the face) to hotel towels (she'd seen a documentary on microscopic bugs that live in linens) to horses (she was afraid they'd think her nose was a carrot and bite it off). Oh, and also ponds ("They're like toilets for birds that never get flushed").

Unlike me, Katie had been popular in high school and, consequently, learning to be self-sufficient had never crossed her mind. If she was too scared to do something, there were always fifty guys around who would gladly do it for her. So she had trouble understanding the growing pile of survivalist books on my table.

"My boyfriend's going a bit nutty," she wrote in her journal right after we started dating. "Something must really be going wrong if he's stocking up on all those books. I hope my hair straightener still works if it does."

At one point in history, almost everyone was a survivalist. They knew how to hunt, farm, fight, and keep themselves and their families alive without the infrastructure and conveniences we have now.

The modern survivalist movement began as a nostalgic yearning for that way of life. Its grandfather was Harry Browne, one of the libertarian authors Greg had recommended at the Sovereign Society conference. Concerned about inflation, the devaluation of the dollar, and nuclear war with Russia, Browne began leading seminars in the late sixties on how to survive an economic collapse in America. He was soon joined by Don Stephens, an architect who gave the movement's followers the name *retreaters*. Worried that cities would erupt into violence in the face of a food, water, power, or other shortage, retreaters advocated building self-sufficient homes and communities in rural areas to flee to.

Then, in the seventies, a man named Kurt Saxon came along. Saxon, an ultra-right-winger who'd grown up during the Great Depression, didn't have much use for people like Browne. At various points, he was a member of the neo-Nazi party, the John Birch Society, and the Minutemen. In 1970, he appeared before a Senate investigation subcommittee after suggesting liberals be bombed. Considering the word *retreater* wimpy, he began popularizing the term *survivalist* in his newsletter, *The Survivor*.

Around the same time, the other fathers of the movement emerged from the right wing. Among them were Howard Ruff, a financial adviser who wrote *Famine and Survival in America* in the wake of the 1973 oil crisis, and Mel Tappan, who published his most famous book, *Survival Guns*, and started a newsletter, *Personal Survival Letter*, in 1977, the year blackouts in New York led to widespread looting and arson.

"If a social breakdown comes, you may be faced with living under primitive conditions for a year, a decade or even the rest of your life," Tappan wrote in one of many essays on his favorite subject, "and your basic life support problems will almost certainly be complicated by encounters with desperate, dangerous

mobs of people who have made no crisis preparations of their own and who are anxious to avail themselves of yours by force. Instead of compromise or improvisation, such circumstances call for the most specialized and efficient arms available."

It was these men who both gave birth to survivalism and gave the term its fearsome reputation.

And so I did what any aspiring survivalist would do: I called them to ask where to begin.

LESSON 32

# LIFE ADVICE FROM THE WORLD'S FRIENDLIEST NAZI

had very little time to learn as much as I could about survival-ism while waiting for the results of my St. Kitts citizenship application. So I decided to start with the man who supposedly coined the word: Kurt Saxon, who was now seventy-six years old. Though he hadn't returned my inquiries when I was a journalist reporting on Y2K, I hoped he'd be more receptive to me as a fellow survivalist.

After tracking down his number, I called and was surprised to hear him pick up on the first ring. Somewhat nervously, I told him I was interested in learning the skills necessary for survival and self-sufficiency. I was sure that after three decades of being the figurehead of the movement, he'd be sick of such requests. Since he'd been a member of so many hate groups, I also assumed he'd be somewhat aggressive, cantankerous, and paranoid.

His response, however, was not one I would ever have predicted.

"Do you feel threatened?" he asked in what may have been the friendliest voice this side of Mr. Rogers.

"I do," I replied.

"You shouldn't." His tone was warm and accommodating,

like he was telling a child not to be afraid of the sound of thunder.

"Why is that?"

"Because paranoia doesn't pay."

"I'm surprised to hear you say that. So you're optimistic about the future of America?"

"Everything is going to hell," he replied matter-of-factly. "I'm a historian. I know what's happening now. I can predict that within five years, four-fifths of humanity will have starved to death."

"If that's true, then why are you saying I shouldn't feel threatened?"

"Well, I don't feel threatened."

"Why is that?"

"Because I live in Arkansas."

This had to be one of the most circular conversations I'd ever had. He was like the Cheshire Cat of survivalism. "I guess that's safer than a metropolis like Los Angeles or New York," I replied, struggling to keep pace.

"Well, you ought to move. I think you'd like it here. For instance, I haven't seen a black person in over a year." I was aghast to hear these words come out of his mouth, despite his background. Perhaps he sensed it in my silence, because he backpedaled somewhat. "This is northern Arkansas. There are blacks in southern Arkansas, but they're nice."

I didn't really know how to respond. I wasn't going to agree, but after years of interviewing musicians, I'd learned that if someone feels you're judging them, they'll never open up to you. So instead of being critical or insisting your views are more correct, you give them what every human being really wants, deep down: acceptance, approval, and understanding. "I'll consider changing my location," I replied, haplessly adding in a struggle to change the topic, "I bet the food is great down there."

"You can't understand how stupid people can be till you move to Arkansas. Anyone who can read without moving his lips here is considered an intellectual. I'm more of an objectivist. Have you read Ayn Rand? *Atlas Shrugged* is one of my best books. I'm kind of a Howard Roark."

I'd started the book three times but never finished it. Yet everyone I talked to, whether PTs or survivalists, seemed to swear by it. It was time to pick it up again. (And when I did, I would learn that Roark was the protagonist not of *Atlas Shrugged* but of Rand's earlier book *The Fountainhead*.)

"I'm neither left-wing nor right-wing," Saxon continued. "Forty years ago, I had a habit of joining nut groups. And you can't find any nuttier group than the Nazis. They were fun. When we disbanded, we formed the Iron Cross Motorcycle Club. And we were the toughest storm troopers out there. We were terrorists. We used terror."

If there is, as paranoid people suspect, an FBI phone-monitoring system that begins recording every time certain words are used in a conversation, Saxon had definitely tripped it by now.

"Occasionally the police would call and tell us there was going to be a hippie bash protesting the war, and we'd go in," Saxon continued. "The police loved to watch because they hated hippies. Still, some of our guys wanted to quit and join the left, because they had better booze, dope, and girls."

In the same patient, grandfatherly voice, Saxon talked to me for another hour, then promised to send a package with back copies of *The Survivalist;* an eighteenth-century compendium of recipes for food and medicine called *The Compleat Housewife;* and all four volumes of his do-it-yourself bomb-making, booby-

trap-setting, chemical-weapon-manufacturing, street-fighting manual *The Poor Man's James Bond.* "My main thing is to save the best of our species," he explained. "And to collect and preserve useful knowledge."

I felt conflicted talking to him. Anyone who joins the Nazi Party for any reason is thoroughly despicable. Yet Saxon seemed so guileless and giving. Then again, if you asked Charles Lindbergh what he thought after meeting Hermann Göring in the 1930s, he would have told you the Nazi commander was a swell guy.

"Are we best friends?" Saxon asked before I hung up.

"Um, I guess so." I didn't know how to reply. Usually it takes a few ball games and nights on the town to become best friends. But not for Saxon.

"Oh, yes," he concluded. "We are!"

And, true to his word, he called me at least once a week after that conversation.

## LESSON 33

# INTRODUCING THE URBAN SURVIVAL KIT

Though it was nice to have the father of survivalism as a friend, it wasn't getting me any closer to learning how to save myself in an emergency. Even his books, when they arrived, were just compendiums of information, invaluable for reference but useless for beginners like myself.

Most survivalists had an advantage over me. Maybe they were born on a farm and knew about crops and livestock. Perhaps their fathers had taken them camping or hunting or fishing. Maybe they'd even had experience at an early age with woodworking, engine repair, or electrical wiring.

But I was raised in urban apartments. Up on the forty-second floor of our seventy-two-story building, we not only had no backyard and no pets, but no fresh air: we were so high up that our windows wouldn't open more than a crack. Repairing things meant calling building maintenance. And dining at home meant that Dad picked up takeout food on the way back from work.

The only useful survival skill I had, thanks to researching a form of persuasion known as neuro-linguistic programming for *The Game,* was the ability to try to talk people with practical knowledge into helping me out. And now I also had a stack of books from Saxon filled with cool plans and diagrams, but no skills or tools to actually use them.

I asked Katie one morning whether she'd learned any survival skills while growing up. "I don't know how to make food," she began. "I can't sew. I can't drive. I can probably talk someone out of killing me. That's the only thing I can do, baby."

Between the two of us, then, conversation was our only survival skill. Perhaps we were a perfect match.

I grabbed a piece of paper and wrote "Survival To-Do List" across the top. Then, since Saxon hadn't been much help getting started, I tracked down Bruce Clayton—or Shihan Clayton, as his website said he was now known since achieving his sixth-degree black belt in Shotokan karate.

Oddly, like Saxon, Shihan Clayton started the phone call by telling me there was nothing to be afraid of. Perhaps it was a reflex from years of being pilloried in the media. "I'm not particularly frightened by the terrorist thing," said the author of *Life after Terrorism*. "They're an out-of-control street gang with grandiose ideas. There are bigger and tougher organizations cruising the streets of Los Angeles."

When he realized I wasn't calling to challenge him but to learn from him, Clayton began to open up. "There are a number of directions problems can come from," he said, finally. "One is, what nasty surprises can al Qaeda come up with in the United States? The second direction is that Iran will get nuclear bombs, though they'll be sorely divided on whether to drop the first one on Tel Aviv or Paris. The third danger is that China and Korea are getting better at building bombs. They can reach the West Coast now, and soon they'll be able to reach all of the United States. So we're looking at another Cold War there. And with rising gasoline prices and ethanol affecting the price of corn, we may be looking at an economically difficult time in the United States."

The fears of Americans change over time. In late 1999, we feared the collapse of our computer system. Then it was terrorist attacks. Then it was our own government. Then it was global warming. Today it's economic collapse. Fear, it seems, is like fashion: it changes every season. And even though threats like terrorism persist to this day, we eventually grow bored of worrying about them and turn to something new. Ultimately, though, every fear has the same root: anxiety about things we take for granted going away.

"The truth is," Clayton continued, "in terms of riots or terrorist incidents, the threat zone is going to be pretty limited. You can walk out of a threat zone, typically. So all you need for the basics are nice shoes and an urban survival kit."

"What's an urban survival kit?"

"Three things: a cell phone, an ATM card, and a pistol."

"I have two out of three. But I've never fired a pistol in my life."

I'd only had firearms in my house once. When I wrote a book with Dave Navarro, documenting the rock guitarist's addiction to heroin and cocaine, I'd removed a handgun and a shotgun from his house because I was worried he'd kill himself. But I had trouble sleeping at night knowing such dangerous weapons were sitting in my closet. Any thief who broke in would know how to use them better than I did. So rather than return them to Navarro—and be responsible if anything happened to him—I gave them away to a friend named, appropriately, Justin Gunn.

"The first requirement in using a gun correctly is not shooting yourself with it," Clayton was telling me. "It's a martial art, and you have to take it seriously. It's not like learning to change a tire. I went out to the Gunsite ranch in Arizona. In a week, they can teach you to be damn dangerous with a gun." He paused and reconsidered. "Actually, the guy who buys the gun and puts it in his pocket, he's dangerous. Gunsite graduates aren't dangerous—they're deadly."

And so I added the first item to my survival to-do list: learn to shoot.

I heard the voice of Spencer Booth gloating in my head as I wrote those three words down. I had crossed over to the other side.

## LESSON 34

# JUST SAY NO TO CHEERLEADING

I need to make two stops," I told the cab driver. "First I have to go to Gun World on Magnolia. Then I'm going to the Burbank airport."

As soon as the words left my mouth, I realized I sounded like a madman. I tried to think of some way to explain this sequence of stops rationally—that I wasn't planning to spray gunfire in the airport but actually picking up a pistol so I could fly to a place called Gunsite in Arizona and learn to shoot. But that sounded just as dubious.

Though I didn't think I'd be allowed to fly with a gun, when I called Southwest Airlines earlier that week I was surprised to discover it was fine, as long as I declared it and kept it unloaded in a locked case in my checked luggage.

The taxi driver, a bald, frowning Armenian in a sweat-stained white polo shirt, took me wordlessly to Gun World, checking me out in the rearview mirror every few minutes. Just in case he drove away, worried that I'd kill him or hold him hostage when I returned with my gun, I brought my suitcase into the store.

I'd first visited Gun World with Katie ten days earlier to buy a Springfield XD nine-millimeter, which Justin Gunn had recommended as a beginner's pistol.

At the counter, a man with black hair, black glasses, a goatee, and a gun thrust down the back of his pants gave me an application that needed to be approved by the Justice Department. It asked typical security questions about whether I'd ever been in a mental institution or had any restraining orders against me. Oddly, it also asked, "Have you ever renounced your American citizenship?"

It seemed strange that this would disqualify someone from owning a gun, as if it were a sign of violent tendencies like a restraining order. Fortunately, it didn't ask if I'd applied for any new citizenships lately. Even stranger, though providing my social security number was optional, the application required me to state my race. It seemed kind of, well, racist.

Next to me, two Polish guys in matching basketball jerseys were looking at .44 Magnums. They'd just turned twenty-one—the legal age for owning a handgun in California. One of them pointed the gun at the mirror, admired himself holding the weapon, then said, "Stick 'em up," and mimed blowing away his victim.

Petrified by this pantomime, Katie asked the man at the counter, "Don't you feel weird selling guns to people? I don't think people should have guns."

"It's in the Constitution, you know."

"They should just be totally illegal. I think only violent people would want to shoot guns."

It was difficult to tell whether the clerk was amused or annoyed.

After I completed the Justice Department form, he explained that due to California state law, I'd have to wait ten days before I could pick up the gun. Then he slid me another piece of paper to fill out. It appeared to be some sort of test.

"What's that?" I asked.

"You have to pass a handgun safety test first."

I hadn't taken a test in years. "Shouldn't I study or something?"

By this point, it was clear that Katie and I were the stupidest people in the store. "Even if you were blind, you'd probably pass," he said.

And he was right:

California Department of Justice
**Handgun Safety Certificate**          **496432**

Strauss | Neil | D
Last Name | First Name | M.I.

B481379 | 10/13/73
CA Driver's License or I.D. | Date of Birth

6/19/08 | 7/19/11
Issue Date | Expiration Date

Signature

While packing for Gunsite ten days later, I dug through my closet to find practical clothing for shooting in the hot desert sun all day and under the cold desert sky at night. I came up nearly empty-handed. For most of my life, I'd bought clothes with the intention of being fashionable and hopefully attracting women. My jeans weren't rugged and durable but slim-legged and waxy, and I wore accoutrements like wrist cuffs and wallet chains, which had no practical use to me—I didn't even carry a wallet. I had the clothing of an urban café dweller, not a survivalist.

Katie tried to stop me as I left for Gunsite that morning. "I just don't picture you holding a gun, babe." She knew me well. "What if you get shot there? Or your gun blows up in your face?"

Those were the same questions I'd been asking myself. "I'll be

okay," I said, as much to convince myself as her. "Why are you always so scared of everything, anyway?" I'd always wanted to ask her that.

"I think I watched too many movies as a kid. And I have a good memory, so I remembered all the scary things I saw. It's a big world out there, and I'm a little person, you know."

Though Katie was reluctant to admit it, movies weren't the only source of her insecurities. Her stepfather used to punish her by making her kneel for an hour holding a box over her head and beat her if she dropped it. Once, on her birthday, he yelled at her for opening her presents carelessly, sent her to her room, and confiscated her gifts. And he constantly told Katie she'd grow up to be ugly, which is probably what led to her inability to leave the house without fixing herself up for an hour in the bathroom.

"When you grow up like that, you spend the rest of your life trying to avoid things that are dangerous," Katie continued, "like dark alleys and cheerleading."

"Why cheerleading?"

"I don't want to end up on the top of the pyramid and fall off and break my bones," she explained as I listened incredulously. "I don't want to be on the bottom, because my back could break. And I don't want to be in the middle, because someone could fall on me." She paused and thought about it a little more. "There's really no safe place in that pyramid, babe."

"But what about bigger threats like terrorism and the economy?"

"I try not to think about those things."

My fears, I realized, were different from Katie's. Mine were forcing me out of my comfort zone; hers were making her retreat. And though it may be tempting to write off her fears as irrational, I later learned that roughly 66 percent of all severe sports-related injuries to female high school and college students are due to cheerleading accidents. And the main cause of those injuries is actually pyramids.

Nonetheless, I needed to find a way to make Katie face her anxieties, because they were holding her back from life and making her too dependent on others to take care of her. Before I could help her with her fears, though, I needed to do something about my own.

After picking up my gun, I resumed my taxi ride to the airport with the nervous Armenian driver.

Fifteen minutes later, with my heart in my throat, prepared at any moment for security to wrestle me to the ground, I walked to the airline counter and repeated the exact words the clerk at Gun World had told me to say: "I'll be checking in a firearm. It's in my duffel, and it's in a locked case and unloaded."

"I'll need to see it," said the ticket agent, a pasty blond woman dotted with moles. She eyed me suspiciously. Maybe I was too nervous.

I repositioned my body in an attempt to block the view of the passengers in line behind me, and removed the case from my duffel. Then I placed it on the scale, unlocked it, and pulled this out:

Nobody panicked, nobody screamed, nobody tackled me. At that moment, if I'd wanted, I could have sprayed bullets all over

the airport. But that would have made me an idiot, and the point of the gun was to protect myself from idiots like myself.

After telling me to remove the magazine, the mole lady asked me to put the gun back in the case and then taped this label to it:

"As a precaution," she added after giving me a boarding pass, "you may want to wait near that door for ten minutes in case TSA has any questions."

When no one appeared after fifteen minutes, I headed through security and to the gate, on my way to Arizona to learn how to kill.

Although a gun can't do much harm in a locked box in a plane's cargo hold, I had no idea it was this easy to fly with a firearm. It was the first time since I began this journey that I discovered a freedom I didn't know I had, rather than a new restriction.

## LESSON 35

# A PHILOSOPHY OF THE SPHINCTER

Make no mistake about what we do here," snarled Ed Head, the operations manager of Gunsite. "This is a fighting school." Sitting in front of him were forty gun-wielding students: soldiers, sheriffs, corrections officers, and me.

"This will be a life-changing experience," he continued. "Most of you are going to leave on Friday a different person. You're going to be better trained than ninety-five percent of all law enforcement officers in the United States." He waited to let the fact sink in. "And you'll develop a different kind of confidence that comes from the ability to take care of yourself."

His last words—"the ability to take care of yourself"—were just what I needed to hear, especially after signing six pages of waivers promising not to sue them if I was accidentally shot in the head.

For the next two hours, the rangemasters, Charlie McNeese and Ken Campbell, indoctrinated us into the mental spirit of battle. "You don't rise to the occasion," McNeese lectured. "You default to your level of training. When the stress hits, you will only be half as good as your best day of recent training."

McNeese appeared to be in his late fifties. With his gray beard, genial Southern accent, and toothy smile, he looked like a dangerous Colonel Sanders. "Excitement won't kill you," he concluded, "but surprise will."

This was the kind of knowledge I needed. It seemed more useful and exciting than knowing the inflation rate and the price of gold and most of what they taught at the Sovereign Society—perhaps because McNeese's words were meant for fighters, not runners, and I still knew nothing about fighting.

After the lecture, we picked up a thousand rounds of ammunition and drove down a dirt road to an outdoor gun range. A sign posted in a shed there read in big black letters, I'D RATHER BE WHACKING TANGOS.

I asked another student, a squat police officer who seemed physically unable to change the shape of his mouth into anything other than a straight line, what a Tango was. "A terrorist," he replied, as if it were obvious (which perhaps it was). He didn't speak another word to me for the rest of the week. I soon earned the derision of the rest of the class by asking the instructor to help me put on my gun belt. It was clear I would be the slowest student there.

I may have had book smarts, but I was sorely lacking in motor skills, practical knowledge, and common sense. In a survival situation, every other person here would most likely outlast me. Perhaps the reason I was so worried about the draft when I was a teenager was that I knew I'd die. If I was on patrol and someone yelled "duck," my head would be the last to go down and the first to get shot.

The only other person struggling with the class was a tank of a woman named Stephanie, a navy officer who was being deployed to Iraq the following month. Her problem was the Beretta nine-millimeter pistol she was using. Not only did it seem to underperform against every other weapon in class, but it was constantly jamming and misfiring.

When I asked why she was using the Beretta, she explained that it's the gun the government issues to most of the U.S. military. In fact, she complained, the government had just bought 25,000 more Beretta M9s. Every time her pistol jammed in class after that, I imagined it happening in an actual gunfight in Iraq.

Other than Stephanie, I was the only other student with a nine-millimeter gun. Everyone else had larger, more powerful .45s, mostly 1911 Colts, which were favored by the late Colonel Jeff Cooper, the severe, charismatic shooting star who founded Gunsite in 1976.

"Why do we carry forty-fives?" McNeese replied when I asked about the gun. "Because they don't make forty-sixes."

At Gunsite, we didn't just learn how to draw a pistol, shoot rapid-fire, quickly clear gun malfunctions, speed-reload, and hit targets at night. We learned the science of killing.

Here, there were no such things as bullet holes. They were called leak points. As McNeese taught us, "The bigger the bullet, the more fluid goes out and the more air comes in."

As Campbell taught us, "When a guy hits the ground, the fight isn't over. He's dropping because the hole in him is causing his blood pressure to drop. But when he's down, his blood pressure will rise again and this means he could still be a threat."

As an instructor named Mike taught us, "If you shoot him in the skull, the skull is a hard thing—the bullet could just deflect and he'll still be standing. You actually want to get him in the eye."

While Mike lectured, I looked around and realized that many of the people there had either killed before or were preparing to kill—and without remorse. In their world, there seemed to be two kinds of people: good guys and bad guys. Good guys should have guns; bad guys shouldn't. And they were the good guys.

It wasn't too difficult to figure this out. One student, a former Green Beret named Evan, was wearing a shirt that read I'M THE GOOD GUY.

Personally, I don't believe in good guys and bad guys, as compelling as those stories may be to children. There are no bad guys—just people who do bad things. Most of them aren't trying to be bad. They think they're the good guys; got stuck in a bad situation and lost control; or have something wrong with their heads that's a product of the way they were raised, the drugs they're using, or the chemistry in their brains. Most of them think they're a hero to someone. And, sadly, they're right—just as every good guy is a villain to someone.

In a Fliesian world, there is no black and white—only gray.

Midway through the week, we were taken to a shoot house. It was a building with multiple rooms filled with human replicas. Some were armed criminals, others innocent civilians. The instructors taught us how to open doors, look around corners, and move through a house where armed intruders are present. Then they sent us into the building one by one to take out the bad guys.

Though it was a simulation, my adrenaline kicked in instantly. I pushed the front door open, then quickly backed away with my gun in the ready position. I saw a man in the corridor in a leather jacket, holding something shiny and menacing in his right hand. I gave him what they called the Gunsite salute: two bullets aimed at the heart, one at the head. When I entered the corridor, though, I discovered that what was in his hands wasn't actually a gun—it was a beer bottle.

For the rest of the exercise, I was shaken. All I could think was, if this had been real life, I would have just killed an innocent person. I suppose that was one of my problems with guns.

"You know what your problem is?" McNeese asked afterward.

"Target identification?"

"Nope." He touched the tip of his index finger to the middle joint of his thumb to make a small circle. I stared at it a second, wondering what it was supposed to symbolize.

"Your sphincter," he informed me.

"My what?"

"Your sphincter. Every time you shoot, it gets like this." He narrowed the circle formed by his thumb and index finger. "It gets tight and itty-bitty. And when your sphincter gets tight, do you know what happens?"

"I can't shoot?"

"You can't even think. When your sphincter gets that tight, it cuts off the blood to your brain. And that cuts off the circulation to your muscles. So your shooting is all over the place." He gave me a friendly smack on the shoulder, hard enough to throw me off balance, as if to prove a point. "If you have a more relaxed muscle, body, and mind, you're going to perform at a higher level than when you're extremely uptight. So you got to learn to relax, son. The sphincter will mess you up."

I watched as the next student entered the shoot house. Since he was a police officer, I figured he'd do better than me. But he shot the innocent man as well.

"You just got yourself a lawsuit," McNeese told him.

"Not if I fix it in the report," the officer joked.

As a journalist, I'd spent most of my adult life interviewing individuals these people considered the bad guys. This was the first time I'd been on the other side of the law—with those whose job it was to uphold it. From what I saw, though, there was little difference between the two groups. It seemed to reaffirm one of my Fliesian beliefs: that people will get away with what they can.

In the evenings, I took an extra class that would earn me a permit to carry a concealed weapon in thirteen states. Accord-

ing to the teacher, a grizzled shooter with a missing finger, courts had affirmed that neither the state nor the police have a duty to protect citizens. "As individuals," he told us, "we have a duty to protect ourselves."

His words echoed the epiphany about being responsible for my own safety that I'd had during the blackout in St. Kitts. Shihan Clayton had definitely sent me to the right place.

Of course, the instructor went a little further, as most people tend to do. After teaching us how to keep from incriminating ourselves after shooting an intruder (hang up immediately after giving the 911 operator your address and don't say a word to police officers without an attorney present), he concluded: "If I kill someone, you won't find me sitting there in my lounge chair with my gun, smiling and telling myself that I done good. I'll be shaking, I'll squeeze a tear out of my eye, and at my age I'll say I have chest pains. The smile will be on the inside."

While listening to him field questions from students about where they were allowed to bring their concealed weapons, I realized most of the people here, even though they seemed tough on the surface, actually lived in fear. They owned guns because they were scared—of gangs in the streets, of robbers in the convenience stores, of burglars in their homes. In their minds, lurking just outside their lives, waiting for the right moment to attack them, were all kinds of men, women, and even animals belonging to a rampant, growing subspecies known as the bad guys.

One of the reasons they preferred .45s was because they were worried a bad guy who was high on PCP might not be stopped by a smaller nine-millimeter bullet. They didn't like being prohibited from carrying guns on planes, in case they were held hostage by hijackers. And they didn't like gun restrictions in national parks, because that left them open to attacks from bears.

In short, they hated any business or law that regulated firearms—because without their firearms, they believed, they weren't safe.

Yet when I asked even the most dogged marksman there if he'd ever used his gun in civilian life, the answer was almost always "no" or "almost." But as easy as it was to find flaws in their logic, ultimately these gun enthusiasts were just like me: they were survivalists. And they wanted to be prepared for every eventuality, no matter how unlikely.

## LESSON 36

# FAMILY-PROOF AMMUNITION AND OTHER TIPS FOR HOME SHOOTING

By the last day of class, I'd taken McNeese's advice to heart. I slowed down, relaxed, and loosened the sphincter muscle. Instead of trying to remember all the details essential to accurate shooting, I simply trusted myself to do it—coolly, slowly, and methodically.

Soon I was pressing the trigger instead of pulling it jerkily, keeping my eye on the front sight of the gun, hitting targets in clusters around the heart and eyes, and making it through entire shoot houses without killing a single innocent. In 1.5 seconds, day and night, I could draw a holstered pistol, aim at a target seven yards away, and shoot it twice in the heart.

It wasn't just a lesson in marksmanship. It was a lesson in life.

"We're in America, we're shooting guns—it's a good day!" Campbell proclaimed as he walked down the range, watching his students blast the hearts out of their targets. "There's nothing better in the world."

In my final shoot house session, I noticed that the first gunman was wearing a bulletproof vest, so I shot him in the head. In the kitchen, I spotted another gunman outside the window and shot him in the heart. And in the last room, there was a target with a gun, but beneath it, he was holding a small child. I asked him to release the child, and when he didn't I shot him in the head—twice.

"I'm impressed," McNeese said. "So you *are* trainable."

A feeling of relief washed over my body. I had succeeded in my mission. I felt like I could protect my home in St. Kitts—and in Los Angeles. At least from small bands of looters who hadn't been trained at Gunsite.

In the process, though, something strange had occurred. I developed a bloodlust I'd never felt before. I actually wanted an excuse to shoot a bad guy, so I could experience what the instructors had talked about.

Guns are designed to kill, and to understand them correctly is to understand killing. Not shooting anyone was like learning to golf on a driving range but never going to an actual course. (Of course, ask anyone about golf at Gunsite and they'll tell you it's "a waste of a perfectly good rifle range.")

I began to understand how armies could so easily indoctrinate new recruits to murder. It's as easy as stereotyping a group of people as bad, then teaching soldiers day after day how to kill them before they kill you. The better the training, the more of a waste it would be not to use it. After all, those who don't know how to fight tend to stay out of fights.

"Be safe," McNeese told us during our graduation ceremony, "and be good to everyone you meet—but always have a plan to kill them."

He then handed us our diplomas:

Gunsite Academy, Inc.
Certificate of Achievement
Having completed the course of instruction in Defensive Pistol
Neil D. Strauss
has achieved the status of  Marksman
in the use of the  Springfield XD 9mm

"I'll tell you something," Stephanie, the student with the government Beretta, told me after the ceremony. "The day I checked in here, I had fired a handgun probably twice before. This course has probably saved my life."

Afterward, we went to Colonel Cooper's house to meet his widow and look through his memorabilia. I sat and talked with McNeese, who recommended some of the other courses. "Don't go into a gunfight with a pistol," he confided, leaning in close. "It's a low-damage weapon. It's what you use when you can't get a bigger gun."

"So what kind of gun do you recommend then?" To me, low damage was getting hit in the head with a Ping-Pong ball, not a .45-caliber hollow-point bullet traveling 835 feet a second.

McNeese suggested a twelve-gauge shotgun with an 18.5-inch barrel. He recommended using slugs for outdoor shooting and, for indoors, birdshot, which is less likely to penetrate a wall and hurt a family member. Apparently, I was going to need heavier artillery for my urban survival kit.

Nearby, a group of students and instructors were making fun of Democrats, gun control laws, and anyone from California. "There's no constitutional amendment that's been more crippled and regulated than the Second Amendment," a competitive shooter was saying about the right to keep and bear arms.

After eavesdropping for a while, I began to realize that all my life I'd been a hypocrite. As a journalist I'd always supported the right to free speech, but been opposed to guns. However, by playing favorites with the amendments, it wasn't the founding fathers' vision of America I was fighting for—it was just my personal opinion.

Though I'd passed the writing and shooting tests at Gunsite, which made me eligible for the permit to carry a concealed weapon, I discovered that I needed to send a set of fingerprints to the State of Arizona to receive it. And, unlike the prints I submitted to get my background check for St. Kitts, these would be kept on file permanently. So I was faced with a dilemma: which was more important, my safety or my privacy?

After weeks of debate, I chose privacy. If the social order ever broke down, no one would be checking for concealed-weapons permits anyway.

Spencer, meanwhile, was solving the same problem the B way. "I've actually been trying to find a podunk place where I can buy the police a squad car in exchange for being named an off-duty cop," he said when I called to tell him about my brand-new pistol skills. "That way, I can carry a concealed weapon anywhere." He hesitated for a moment. "I want to have it ready for the next fiscal crisis."

This wouldn't be my last gun class. I would eventually purchase and learn to use a Remington 870 Wingmaster shotgun with a ventilated rib barrel and a model 700 rifle with a Tasco Super Sniper scope.

Thanks to Kurt Saxon, Mel Tappan, and Bruce Clayton, I'd become a gun nut. I'd become one of the guys I would have been too scared to hang out with on the millennial New Year.

# ⚠ WHY YOU SHOULD THINK OF THE ZOO AS A RESTAURANT

There is a language of survival.

WTSHTF is short for When the Shit Hits the Fan. And, as disastrous as that may sound, it's not nearly as bad as EOTWAWKI—the End of the World as We Know It.

Bugging out is slang for leaving your home to go somewhere safe. To do so, you'll want a bug-out bag (or BOB) full of survival supplies for the road; a bug-out vehicle (or BOV), which will get you out of the impact zone and through traffic as quickly as possible; and a bug-out location (or BOL) stocked with enough provisions to get you through whatever crisis is occurring.

So, in short, WTSHTF, you're going to want a BOB to bring into your BOV to go to your BOL, where you'll pray it isn't the EOTWAWKI.

I learned all this on the Survivalist Boards.

When I returned home from Gunsite, I noticed that most of my friends grew fidgety whenever I began discussing my quest for a passport, a Swiss bank, and bigger and better firearms. Either they thought I'd gone off the deep end or they wanted to talk about something more directly related to their interests— like whether *Rock Band* was better than *Guitar Hero*.

This was before gas prices soared, banks collapsed, and my

friends began feeling the terror in their wallets. Most people tend not to care about political, economic, and ecological problems until they're personally inconvenienced by them.

Thankfully, there's this thing called the Internet, where individuals who feel alone in their predilections can discover a community of people who share their interests. If Armin Meiwes, a cannibal in Germany, could use the Internet to find two hundred people volunteering to be killed, cooked, and eaten, surely I could find a community of survivalists to give me advice.

All it took was one search for *survivalist forum* to find survivalistboards.com, which soon became a daily obsession. The Survivalist Boards were a treasure trove of postapocalyptic suggestions, ranging from hunting for food at the local zoo if game becomes scarce to making a gun out of galvanized plumbing pipe, a dowel, a nail, duct tape, and a small piece of cardboard.

I read the boards for days, trying to separate the useful, practical advice from the black-helicopters-following-me paranoia. I didn't know when—or whether—my St. Kitts citizenship would come through, so I wanted to train as quickly, intensely, and efficiently as possible. In addition, because I didn't plan on living in the Caribbean full-time, I needed to be prepared for the more likely scenario that something would go wrong while I was still in Los Angeles.

As long as I was joining hidden communities, I also signed up for a perpetual traveler mailing list called PT-Refuge. Survivalists understood handheld tools like guns, knives, archery bows, fishing poles, and can openers. But they had almost no interest in nontangible tools like passports, trusts, LLCs, and numbered accounts. Perhaps they didn't think these things would still be around WTSHTF. Or, more likely, judging by some of the survivalists I corresponded with, they clung fiercely to their own conception of America, and a SHTF scenario was also an oppor-

tunity to form a militia and remake the country according to their own ideals.

I called Spencer to tell him to check out the Survivalist Boards. He was in St. Kitts with his brother and his parents—who, unlike my family, supported his endeavors. They were getting citizenships too.

"I knew you'd catch on eventually," he told me.

"Thanks, I think."

"These are the kinds of things you should be considering," he continued as he scrolled through the posts. "The way I see it, there are only two situations we need to be prepared for."

Since I'd discovered the boards, I'd been thinking the same thing. "Bugging in and bugging out?"

"If that's what they call it, yes. First, you need to have enough supplies at home so you can survive with no help from anything or anyone in the outside world. Next, you need an escape route in case you have to leave your home. And that escape route should lead you to a safe retreat."

"Like St. Kitts."

"Once we get our passports." He sighed. He'd been feuding over contract details with the developer of the property he wanted to buy, so he hadn't even filed for his citizenship yet. "Our biggest challenge is going to be just getting out of the city, because traffic will be bumper-to-bumper."

"I'll tell you what. Have a good time in St. Kitts. I'll look into ways to bug out when the shit hits the fan."

I was already talking like a hard-core survivalist.

# MOTORBIKES OF THE APOCALYPSE

After getting off the phone with Spencer, I made my first post on the Survivalist Boards—a plea for advice on how to escape the city WTSHTF.

Most of the experts advised moving out of the city immediately. But just like the decision not to turn my fingerprints in to the government, I wasn't ready to compromise my freedom and enjoyment of life for security. A second citizenship added options and opportunities to my life; living next door to Kurt Saxon would only remove options and narrow possibilities.

Others responded with vivid scenarios worthy of a Hollywood blockbuster. One poster, who went by the name Dewey, advised, "The best thing, given your situation, is to: arm yourself to the teeth to fend off looters and gangs, build yourself an excellent BOB, and keep a month's worth of rations at your residence to weather out the initial exodus of sheeple and eventual gang wars. I'd lay low, let all the gangs kill each other off first (expect that within 2–3 weeks), then prepare to hoof it. Thus, an excellent BOB is essential, as is carefully selected firearms. You're definitely going to need a semi-automatic rifle with high magazine capacity."

Others gave less morbid advice that I never would have come up with myself. "Might I suggest a different path?" Lasercool wrote. "If escaping beforehand is not an option, you might want to get involved with the local disaster management organizations run by the government. Here in Miami, the police run a CERT team—Community Emergency Response Team. You can get decent training, a snazzy vest, and most of all contacts within law enforcement. You'll be seen as one of the 'good guys,' and often that can make the difference between getting through a roadblock and not getting through."

Finally, a handful of others on the forum insisted that I purchase a motorcycle with saddlebags and study local trail maps so I could escape via isolated mountain roads instead of crowded, chaotic highways.

The only problem was that I didn't know how to ride a motorcycle. I'd tried once before during an interview with Tom Cruise at a motorcycle raceway. But it ended in disaster: while trying to brake on a turn, I wiped out on his expensive 955cc Triumph bike. Though he didn't seem to mind the damage to the motorcycle, I was humiliated.

Fortunately, the survivalists were helpful and thorough. They suggested the best place to train: the Motorcycle Safety Foundation. They suggested the best survival bikes to buy: the rugged Rokon Trail-Breaker used by U.S. Special Forces in Desert Storm, the versatile dual-sport Suzuki DR-Z400SM, and the Russian Ural, a sidecar-enhanced motorcycle built to battle the Nazis. And they suggested the biggest hazards to watch out for: a saboteur stringing piano wire across the road to decapitate me, an attacker lurking around a corner wielding a two-by-four, and a motorist shooting me in the back to get a free motorcycle.

I was impressed by their imaginations. They reminded me of Katie and her movie-inspired fears. Maybe she'd make a good survivalist after all.

Since the risks seemed improbable, I signed up for a course with the Motorcycle Safety Foundation and looked into the

Rokon Trail-Breaker. With a two-wheel drive system, wide tractor-like tires, and hollow aluminum wheels capable of storing gas and water, the Rokon was a survival machine, able to ride through mud puddles and streams two feet deep, over mountains of rubble, and through snow-covered fields. While cars were stuck bumper-to-bumper on the freeway, I'd be able not just to weave around them, but to cruise along the median, the shoulder, the embankment, and even, if necessary, over the tops of abandoned cars.

For the next step in my evacuation route, I did something Spencer had advised when we first met: I researched flying lessons.

I couldn't believe I was about to take my pursuit of offshore safety this far, but Spencer was right. If I was going to have a compound in St. Kitts, I'd need a way to get there. Though the survivalists recommended using a kit to build my own ultralight plane, my tool skills at the time prohibited anything with words like *kit, build, model,* or *construct*—unless they involved small interlocking rectangular blocks or cheap Swedish furniture.

"The most unforgettable day of your life will be the day you take your first solo flight," Taras, my instructor at Justice Aviation, told me when I signed up for pilot class. "The second-most unforgettable day of your life will be the day you take your first solo trip somewhere."

As I sat through his orientation, it occurred to me that flying lessons would be a long and costly commitment. On the positive side, spending time at the airport would enable me to make friends who flew planes and worked in control towers in case I needed them. On the negative side, a single-engine plane on a day with a good tailwind would only fly as far as Santa Fe on a tank of gas.

I decided to take a few lessons anyway, until I came across

a better option. Though a single-engine plane wouldn't get me to St. Kitts, at least it would get me across the border.

Before I could start flying, however, I needed to give Taras a copy of my driver's license and proof of my U.S. citizenship, such as a passport or birth certificate. Between my flight lessons, gun purchases, and suspicious Internet searches (at least the ones I'd made before taking Grandpa's anonymity advice), I was pretty sure I'd caught the attention of the government. I'd recently read that James Moore, the coauthor of a negative book on Karl Rove called *Bush's Brain,* had been added to the no-fly list. Hopefully, I wouldn't be next.

The tools we need to protect ourselves, I realized, are nearly identical to those that others are using to kill us. Perhaps the only difference between the good guys and the bad guys then is intent. Because by not trusting the state, I'd made myself indistinguishable from its enemies.

# BATHROOM TIPS FOR COMBAT SOLDIERS

Motorcycle school was just as humiliating as gun school. Not only was I inexperienced, but I was one of the few people in class who'd never even driven stick shift before.

Perhaps I wasn't so much getting prepared for hard times as I was catching up to normal people. Driving stick, shooting guns, farming, and using tools are things every man should know. But my friends and I had never bothered to learn these basic skills because everything had always been handed to us. Public transportation, fast food, the Yellow Pages, and Craigslist had made them seem unnecessary.

Just as I squeezed triggers too hard, I throttled the bike too much. When weaving around cones, I inevitably crushed them. And during a stopping exercise, I braked so suddenly that the rear wheel locked and the bike tipped over.

It wasn't until the last day of class that I realized what was holding me back: my sphincter. I was worrying too much—about braking, throttling, balancing and releasing the clutch at just the right times while under the scrutiny of the instructors. And the more I thought, the worse I rode.

As I approached a row of cones, I took a deep breath and decided to trust my instincts instead of thinking so hard about trying to do everything perfectly. This time I carved around the cones gracefully, without hitting a single one. Yet when it came time to repeat my performance during the final driving exam, my sphincter shrunk back to the size of a needle's eye and I didn't even come close to passing. To truly learn to survive in stressful situations, I'd need to take the motto of the Survivalist Boards—"endure—adapt—overcome"—to heart and learn to relax my sphincter permanently instead of tightening it whenever confronted with a challenge.

Or, as security consultant Gavin de Becker put it less scatologically in his book *The Gift of Fear,* I needed to learn to disengage my logic brain and engage my wild brain, which instinctively guards and protects.

To do this, I bought a book an instructor at Gunsite had mentioned: *On Combat* by Dave Grossman, which deals with the body's stress response during battle. In fight-or-flight situations, I learned, one of the first things the body does is give up sphincter and bladder control (hence the phrase scared shitless) so it can commandeer every bit of available strength for battle. That's why some soldiers actually make it a point to go to the bathroom before combat.

So maybe McNeese was right about the sphincter. It is for times of peace, not war.

Before retaking the motorcycle test, I decided to get a Rokon so I could practice on my own:

To help offset the cost, I sold some rare books, CDs, and an old computer. Since I'd gone into survival mode, my intellectual pursuits had begun mattering less to me than physical ones.

One afternoon, as I was practicing to retake my motorcycle test, Katie and her sister pulled up outside my house. Because she was too scared to drive, Katie was almost completely dependent on her sister to get around. However, her sister resented the obligation and frequently failed to pick her up. So Katie had recently been forced to drop out of college for poor attendance and fired from her department-store job for missing shifts.

When she saw me on the Rokon, her brow knitted and her fear reflex kicked in. "I don't like it," she said. Her fingers instinctually rose to stretch her forehead taut in an attempt to prevent wrinkles from forming. "It's dangerous. I don't trust anything with less than four wheels."

For once, I agreed with her.

I drove up the street, followed a bend in the road near a neighbor's house, and tried to practice stopping. As I squeezed the brake, the wheel began to slide out from under me.

I had forgotten the one lesson I'd learned from my wipeout on Tom Cruise's bike: avoid braking when turning.

I panicked and, as I released the brake, accidentally throttled the bike. Instantly, the Rokon accelerated in an arc, rolled up four of the neighbor's stairs, and careened off the ledge to the left, colliding midair with a row of garbage cans. The bike, the cans, and my ribs all clattered loudly to the pavement.

My first thought was to marvel at how easily and smoothly my survival bike had climbed the stairs. My second thought was that McNeese and Grossman were right: because I hadn't learned to stay cool in regular life, I'd panicked in an emergency and hurt myself.

Only then did I roll up a pants leg and lift my shirt to inspect the damage. My lower body was a red skid mark oozing blood.

I was turning out to be a shitty survivalist.

In the last month I'd bought guns, started riding motorcycles, and signed up for flying lessons. In my quest for safety, I was undertaking the most dangerous things I'd ever done in my life. The chances of being injured in a terrorist attack or civil unrest were far slimmer than the chances of killing myself in a motorcycle or shooting accident.

Survivalism, I realized, is not about staying alive. It's about choosing how you die.

## LESSON 40

# NEVER DRINK
# FROM THE BOWL

K evin Mason, a fireman with Fire Station 88 in Los Angeles,
paced back and forth, agitated, in a back room of the First
Presbyterian Church in Encino. He was tall, with gray hair and
the hardened humor of someone who'd seen people die in his
arms. "If there's a big disaster," he was saying, "you cannot expect
assistance for how many days?"

"Three to five days," forty people recited in a staggered response.

"You cannot count on us," Mason continued. By us, he meant
the fire department, the police, the ambulance companies, the
national guard—anyone. "So who's going to get you when there's
an emergency?"

"Nobody," the class thundered.

"Nobody's coming to your aid in a disaster," Mason said, drill-
ing the point into the head of every student, businessperson,
housewife, and grandparent in the room. "You have to be independent."

This was the CERT class the survivalists had recommended.
And it had already taught me that my expectation that the government would save and protect its citizens after Hurricane Katrina was unreasonable. According to Mason, the federal plan
was and always had been: let the mess happen and hope the peo-

ple take care of themselves. Then come in, scoop up the survivors, and help the community recover.

The CERT program was originally developed by the Los Angeles Fire Department in the mid-eighties to help citizens and communities become more self-sufficient in the event of a disaster. After the Federal Emergency Management Agency caught wind of CERT many years later, it took the program nationwide.

The more I learned, the more surprised I was that I hadn't made any disaster preparations earlier. Up until then, I'd been so focused on man-made and political catastrophes that I'd given little consideration to nature's many forms of population control. In Los Angeles, an earthquake or wildfire could easily devastate the city, while in St. Kitts a hurricane could decimate the island. In fact, according to the Uniform California Earthquake Forecast, California had a 99.7 percent chance of experiencing a severe earthquake rating 6.7 or higher on the Richter scale in the next thirty years.

"You have to get that through your head," Mason was saying. "Every year in the country, there are twenty-five thousand disasters. When's our turn?"

"Anytime," the class recited.

"That's right. It's sooner than you think. The San Andreas Fault is overdue. So who's responsible when that happens?"

"We are."

"Right. And"—he smiled wryly at his students; they knew the line coming next—"it's not looting if you leave a note."

This was class number three of seven. When it was all over, I would be issued a green CERT vest, a CERT helmet, a patch, a diploma, and a card identifying me as a CERT battalion member. I had originally taken the class solely to obtain these signifiers of authority, hoping they would make bugging out easier. When escaping through the mountains on my Rokon Trail-

Breaker in my CERT uniform, I figured I'd be able to get past roadblocks by saying I was doing relief work.

But I was surprised to discover that the classes were actually valuable. Each one had the potential to save my life. Furthermore, it was reassuring to hear someone with experience, authority, and government credentials justify the concerns I'd been having for the past few years. They didn't seem like paranoia anymore. They seemed like common sense.

Our first class was about general emergency preparedness. The average person, I learned, needs a gallon of water a day to survive—but after a massive earthquake, Los Angeles could be without water for up to thirty days while workers fix the pipes.

So how do you get the gallons you need, especially if you haven't stored any water? Mason taught us there are forty gallons of drinkable water stored in most home heaters, as well as gallons more in toilet tanks. But, he warned, in his world-weary, people-are-idiots voice, "Never drink from the bowl."

"If you live next door to Noah and he's building an ark, you best be building one too," he told us while listing items to stockpile. So many of the tips I'd been given by neo-Nazis, gun nuts, and fringe weirdos were actually the same things the government recommended doing. The system actually wanted and encouraged us to be prepared to live without it. It was designed for its own obsolescence.

Between classes, drawing on my survivalist books and the Survivalist Boards for additional guidance, I began stockpiling. I stored ten gallon-sized jugs of water in the garage, taking care not to put them on the concrete floor, which would react with the plastic and contaminate the water. I bought a two-thousand-watt Honda generator, along with a hand-cranked generator. I filled five two-gallon containers with gasoline, then added Sta-Bil to prevent the gas from breaking down. And I tracked

down lanterns, kerosene, forty-four-hour candles, hand-crank-powered flashlights, rechargeable batteries, strike-anywhere matches, water purification tablets, and first aid supplies.

My goal was to eventually conduct what Bruce Clayton in *Life after Terrorism* called a three-day test, which meant shutting off all utilities and living off my stockpile for three days to see what supplies I'd overlooked. Afterward, I planned to bring a duplicate set of provisions to St. Kitts.

In class two, we learned about different types of fires and the various extinguishers designed to put them out. We were taught the five-second rule: if the fire doesn't get smaller or go out within five seconds, evacuate immediately. Then we were taken outside, handed an extinguisher, and given the chance to put out an actual fire, which taught me something very obvious I'd never thought about before: always aim the nozzle at the base of a fire, where the flames are interacting with the fuel.

And now I was in class three. Today's topic: handling mass casualties.

"This is the way we look at disasters: we'll get to it," Mason said. "Katrina came and went. We'll get to it."

I was surprised by the casual way he treated disasters, as if they were television shows that reran every night rather than tragic historical anomalies that destroyed lives, families, communities, and even generations. He was more of a Fliesian than I was. I, at least, lived in a world of order that was occasionally disrupted by chaos. As a fireman, he lived in a world of chaos occasionally interrupted by order.

"The goal is not to save everyone," he instructed us. "It's to help the greatest amount of people in the least amount of time. So you have to prioritize. You have to sort."

That day, we learned START, which stands for Simple Triage and Rapid Treatment. If we were the first to arrive at a mass-casualty incident, our job was to sort through the victims swiftly and mark them—with a Sharpie or lipstick on their forehead if triage tags weren't available—as one of four types of casualties: immediate, delayed, minor, or dead. This way, when the professional rescuers came, they'd know who needed treatment and transportation to a hospital first.

After triaging, if rescuers still hadn't arrived, Mason instructed us to create treatment areas for the victims. "Have bystanders get you medical equipment. Check people's trunks for emergency supplies. If there's a nearby grocery store, have volunteers get whatever they can carry. And remember . . ."

The whole class repeated in unison, "It's not looting if you leave a note."

Despite my paranoia, I didn't think I'd ever have to go to this extreme. But in less than a year, I'd be living this scenario.

"Find somewhere cool, like the ice chest of a 7-Eleven or an underground parking lot, to store the dead," he continued. The class seemed to get more macabre each week. "The coroner will love you if you wrap them with plastic. And don't forget to provide some sort of security, because people will strip the dead of their belongings." There was no irony or humor in his voice, just a lack of faith in human goodness.

History is full of tales of people who behave altruistically in disasters, even when resources are scarce. Some will sacrifice their lives to save someone else or give away their last sip of water or share their shelter with a destitute family. But history is also full of people who behave viciously in disasters and will betray, torture, and kill fellow victims to save themselves. It has yet to be proven which group has a better chance of living, but there's no doubt which group is better able to live with themselves afterward.

"Who do you think my best friends in the neighborhood are?" Mason was asking. "The plumbers, contractors, and carpenters. I have a Christmas party every year and I always invite those guys. Why? Because when it hits the fan, I'm going to need them."

Apparently Mason was as thorough a survivalist as the guys on the Internet forum.

I was disturbed to discover that almost every bit of folk wisdom I'd previously learned about surviving natural disasters was wrong. I'd always thought that in an earthquake, for example, the safest place to be was in a door frame. But Mason taught us that's true for the small number of houses made of adobe or un-reinforced masonry. In the majority of homes, not only is the door frame one of the weaker parts of the structure, but if you stand inside it you're likely to get knocked out by flying debris or the door itself.

The safest thing to do during an earthquake, he said, is hold fast under something sturdy like a table—after checking to make sure there are no windows nearby or heavy objects hanging over head. As for a hurricane, if you're unable to evacuate, close and brace all exterior and interior doors and windows. Then lie on the floor under a sturdy object in a small windowless interior room or closet on the lowest level of the house.

Before heading to my fourth CERT class, I tried to talk Katie into coming along. "If you know what to do when bad things happen, then maybe you won't be so scared of them anymore," I told her.

"I don't know. I think if something bad's going to happen, it's going to happen." She muted the movie she was watching, *Interview with the Vampire*. "If there's an earthquake, the ground will be shaking underneath you, and you'll fall over and hurt your head. There's not much you can do about that. You can't walk around with a helmet all day long."

She didn't even have to pause to invent this scenario. I had no clue where some of her ideas came from, because they were too absurd for movies. Perhaps Tom and Jerry cartoons.

I tried to reason with her. "But what about fires? Wouldn't it be useful to know how to put them out?"

But she turned out to be more reasonable. "I don't think I need to learn to put out a fire, because I don't want to be anywhere near a fire. If I see one, I'll just avoid it."

Our approaches, I realized, also reflected the way we dealt with problems in life.

In that week's class, we learned how to stop bleeding, splint broken bones, care for burns, and preserve amputated body parts. The lesson was so useful—not just for survival, but for everyday life (minus the amputations)—that immediately afterward I signed up for a slightly more thorough course with the Red Cross to receive these:

| | |
|---|---|
| **American Red Cross**<br>*Together, we can save a life*<br>✚ | This recognizes that<br>**Neil Strauss**<br>has completed the requirements for<br>**Standard First Aid**<br><br>conducted by<br>**ARC of Greater Los Angeles**<br>Date completed<br>The American Red Cross recognizes this certificate as valid for   3   year(s) from completion date. |

In the next class, we learned the basics of search and rescue. We were taught how to assess whether a building is safe to enter after an earthquake, how different structures are affected by seismic shifts, which parts of a damaged building are most likely to contain victims, and how to use cribbing—small blocks of wood—to lift rubble weighing hundreds or even thousands of pounds off trapped victims.

"Who are the rescuers?" Mason asked afterward.

"We are."

"That's right. We're not going to be able to run the city. If something goes wrong, we'll be looking to you to run the city. Why do you think we give you that green vest with the City of Los Angeles logo on it? You are the trained personnel in a disaster. Because we're not coming. We've told you that for five weeks."

I looked around the room. Half the people seemed like they couldn't climb a staircase without getting winded.

"That C on your vest is for community," he continued. "You don't want to be the lone wolf. What happens to the lone wolf? He gets picked off."

I reflected on his words, because unlike most people in the

class, I wasn't there to be part of a community team. I *was* the lone wolf.

Fortunately, Mason had second thoughts. "Actually," he added, perplexed, "I guess the lamb gets picked off. I don't know what happens to the lone wolf."

The next class, in which we role-played various disaster scenarios, was taught by a different fireman, Sergio Mayorga. Though he was less cynical than Mason, he was no less apocalyptic.

"Next week, after my terrorism lesson," he warned as class ended, "you're not going to be able to sleep for two weeks."

# LESSON 41

# PROTECTING YOURSELF FROM SNIPERS, DIRTY BOMBS, AND SALAD BARS

Why does terrorism exist?" Sergio Mayorga asked.

It was my final CERT class. A box on the table behind him contained forty much-coveted CERT uniforms. At the end of the lesson, one of them would be mine.

"Envy?" An older woman in front of me responded.

"Anger?" an Asian businessman asked.

"It exists," Mayorga said, "because it's faster than the political system." He paused to let the answer sink in. "It's cheap, mobile, low-tech, and deniable. Most of all, it exists because it works."

His argument made sense. Saying "Please keep off my lawn" is a much less effective deterrent than, for example, aiming a shotgun at the trespasser and saying, "Get off my lawn or I'll blow your fucking brains out." And then doing it. No one will set a toe on your lawn again.

"Do you know what B-NICE means? You can use that acronym to remember the different types of terrorist threats: biological, nuclear, incendiary, chemical, explosive."

241

This decade, the James Bond supervillain—a non-national rogue with a heavily financed conspiracy to destroy the world with no material benefit to himself—had actually stepped out of the realm of fantasy and into reality. As John Robb put it in his book *Brave New War,* thanks to "the leverage provided by technology," eventually "one man" will be able "to declare war on the world and win."

"Though fires are the cheapest and most common form of terrorism," Mayorga continued, "the one with the higher risk factor is biological, because it will take a week or two before people start having symptoms and figure out what's going on. The water supply is one way for them to spread something. Another way is to use aerosol and spray the vegetable aisle in a grocery store or the salad bar at Souplantation."

The room was silent as he spoke. Though wildfires and earthquakes were more likely, terrorism was more chilling.

He said, in the event of a nuclear attack, we should hole up in either a basement or the third-from-the-top floor of the highest building around, then cover all cracks and openings with plastic and duct tape.

I opened my class binder and started taking notes.

He said, a dirty bomb is more likely than a nuclear attack because it's easy to make. All someone has to do is build a pipe bomb, then fill it with radioactive material, which can be stolen from any doctor's office with an X-ray machine.

I was writing as fast as I could. I recalled the instructions for making a rudimentary pipe bomb from one of Kurt Saxon's books—just stick a plastic baggie in a pipe, fill it with safety-match heads, add a piece of string as a fuse, and cap both ends. Even I could do that.

He said, if you're in a shopping mall and someone sets off a

dirty bomb, you have roughly thirty minutes to decontaminate.

Why was I taking notes so furiously?

He said, cover your nose and mouth immediately with anything that wasn't exposed.

Was this ever really going to happen to me?

He said, never use a cell phone or radio within a thousand-foot radius of an explosion in case the frequency sets off a secondary device.

How many dirty bombs had even been set off in American shopping malls?

He said, if you rush toward the exits with everyone else, you risk not only being trampled but also getting injured by a secondary device planted there. Instead, head to the bathroom, find the sprinkler, bust it open, and decontaminate yourself.

And how often was I in a shopping mall? Almost never.

He said, if you have to remove contaminated clothing, don't pull it over your head. Cut it off.

Yet I continued scribbling away, because I was haunted by the same demon everyone else in class was haunted by: Just in Case.

I was reminded of a book I'd recently read on the Inquisition. For five hundred years of Western history, the Catholic Church was much more extreme than today's Muslim fundamentalists. Its armies killed all the heretics they could get their hands on, wiping out entire religions by torturing, slaughtering, and burning alive nearly every devotee. Like the fatwas of hard-line Muslims today, it even wrote edicts prescribing these genocidal terms. And because of these bloody measures, Christianity dominates the West today. In history, whether you look at the manifest destiny of America or the city-demolishing bombings that ended World War II, winning is killing.

He said, if you're ever in a public area and someone starts spraying gunfire, get down on your stomach with your feet pointed toward the attacker, your face positioned away, and your hands covering your head. This way, it's less likely that the bullets will hit your vital organs.

Scribble. Scribble. Now this seemed more likely to happen. Slightly.

He said, if you notice an absence of birds and mosquitoes, or you see people with drool, tears, and snot pouring out of their faces, it's most likely a chemical attack. Cover your nose and mouth, head upwind, and use the rule of thumb—which means getting far enough away from a chemical incident that when you stick your thumb out and squint at it, your digit covers the entire scene.

Why hadn't anyone told me this before? I was glad I had to wait for my second passport. Otherwise I might never have taken the time to learn about the dangers I was trying to protect myself from.

He said, if chemicals from the attack get on your skin, rinse them off for twenty minutes with cool water while cleaning with soap and bleach. Don't use warm water because it will open your pores.

Fuck calculus. Fuck trigonometry. Fuck long division. This was cool knowledge, useful knowledge, life-saving knowledge. Even if my chances of using it were as likely as having to use the law of quadratic reciprocity in real life, at least it seemed practical—at least it slew one head of the many-headed demon Just in Case.

Afterward, with no test and little ceremony, we collected our uniforms and credentials—including this certificate, auspi-

ciously signed by the mayor, stating that I was now qualified to serve the city in the event of a disaster. As a committed non-joiner of teams and committees, it wasn't the kind of honor I'd ever expected to receive:

FIRE DEPARTMENT
**Certificate of Completion**
be it hereby certified that

*Neil Strauss*

*has completed the Fire Department's official*
*COMMUNITY EMERGENCY RESPONSE TEAM*
*training. In the event of an earthquake or other disaster,*
*this member is qualified to serve as a civilian member of the*
*Community Emergency Response Team Program.*

A few weeks later, I received a message that ten CERT members were needed at Fire Station 88. We were asked to bring our vest, hard hat, gloves, goggles, and a flashlight. I thought this might be a chance to put my newly learned skills to work.

Instead, it was just a media opportunity. For KTLA News cameras that night, we demonstrated how to remove a trapped victim from under a slab of concrete using cribbing. I was the safety officer. If you look at the still carefully, you'll see me on the far left of the frame—the skinny lone wolf in his hard-earned green helmet, with his foot up on a slab of concrete and his hands protecting his crotch from some unknown threat.

As the safety officer, it was my job to make sure no one was injured. I accomplished this task by watching everyone else do all the work.

Some call it laziness. I call it a survival skill:

# THE GLASS OF WATER WORTH ITS WEIGHT IN GOLD

W hat do you need?

It's great to have Louis Vuitton Cherry Blossom Papillon Murakami purses and Vacheron Constantin Patrimony watches and THX-certified mega-multiplexes with twelve different blockbusters to choose from. After all, most Americans, if given the choice between having the right to vote or the right to see whatever movies they want, would choose the latter.

I probably would. There are a lot of really good movies out there, and not a lot of really good political candidates.

But these are just wants. Happiness, friendship, education, entertainment, success, family, respect, freedom of speech—those are wants too.

Our needs are actually very few. Here is all we need:

1. Safety and health: In short, the confidence of knowing we'll wake up each day in satisfactory working order, without a knife held to our throat or a hurricane tearing our roof off or a disease attacking our body.
2. Shelter and warmth: According to the rule of threes, though we can live without water for roughly three days

and without food for some three weeks, it can take just three hours to die of exposure.

3. Food and water: No explanation necessary.

That's it. Call it the triangle of life. When it comes down to it, every one of us would sacrifice our freedom to fulfill any one of these three needs if they were lacking. And all the aspirations people have—to become rich or famous or simply better than their neighbor—would instead be focused on obtaining a glass of water, a warm place to sleep, or an antibiotic to fight an infection.

There's a ceiling over our heads that we're all striving to reach. And right now, despite the political and economic problems in the country, that ceiling is very high and almost anything is possible. That's why people like Tomas immigrate to America.

Survivalism is preparing for the day when that ceiling is lowered, almost to the floor, and success is simply a matter of not dying. And just because I was learning to ride a motorcycle and fly a plane, just because I was stockpiling supplies and preparing an escape route, just because I knew how to crib and triage, didn't mean I'd truly know how to survive WTSHTF.

So I looked to the people who were the experts at throwing shit in the fan: the military.

# HOW TO MAKE A MAN TELL YOU ANYTHING

There's a program in the military that supposedly turns boys into men. It teaches downed air force pilots to survive behind enemy lines and captured soldiers to resist interrogation.

It's called SERE, and it stands for Survival, Evasion, Resistance, and Escape. It's supposedly hard-core, and hard-core was what I needed if I wanted to learn to fend for myself in both America and St. Kitts.

So I called Evan, the Green Beret I'd met at Gunsite, and asked if they let civilians into SERE.

"SERE's gay," Evan replied instantly. "They don't teach you how to survive. They teach you how to die slowly."

"What do you mean?"

"They basically ask who wants to be an instructor, and whoever says yes gets to teach. You're better off reading a book. The only part that's kind of interesting is the interrogation."

During his mock interrogation, Evan said he was imprisoned, tortured, and waterboarded. At night, his captors—fellow soldiers playing a role—blasted a Nazi SS march at deafening volume over and over, until he was driven nearly mad with sleep deprivation.

"They got pretty sadistic," he continued. "But people underestimate their ability to deal with pain. What got me wasn't the torture, but the psychological stuff. When they put my balls on a table and brought out a hammer, I just broke down."

"So maybe it's worth doing SERE just for the experience?" I prodded. "I mean, minus the balls thing."

"Nah, the rest is pretty lame. The guy you really want to go see is Tom Brown."

"Who's that?" I'd never heard of him before.

"Tom Brown is the real deal. He teaches the guys who teach the marines how to survive, if you know what I mean. All our snipers secretly go to that course. He's a civilian, but he's worked for OGA"—other government agencies. "If you want to learn how to live in the wilderness with nothing but the clothes on your back and a knife in your hands, go see Tom Brown. Fuck SERE."

Those four words—"go see Tom Brown"—were among the best advice I'd gotten in my life.

But there was a problem: Brown's Tracker School was in a campground in the woods. And while camping may be a vacation to many people, it had always been like waterboarding to me. In Chicago, there wasn't much nature around the apartments where I grew up. In the desiccated grassy areas that the city called parks, our main recreation as twelve-year-olds wasn't camping or joining the Boy Scouts, but a game we called jump-the-bum. The object was to leap over sleeping and passed-out homeless men in the park—either alone, in tandem, or en masse—without waking them.

We usually won.

The only other nature experience I had was at overnight camp, but both times I went camping, it poured rain and the cheap tents leaked until we were wet, cold, and fed up. Maybe it was nature's revenge for our jump-the-bum games.

After my camping washouts, I vowed never to set foot inside a

tent again. After all, who needed tents when there were perfectly good hotel rooms nearby with movies on demand and daily maid service?

If I wanted to be a survivalist, however, I'd need to break that vow.

To find appropriate gear, I called Justin Gunn, who, besides holding Dave Navarro's firearms for me, had been trying unsuccessfully to get me into one of his other hobbies, ultralight backpacking. The idea was to go hiking and camping with just a few pounds of equipment—most of it designed to weigh almost nothing and serve a variety of functions. Walking sticks doubled as tent poles, jackets converted into sleeping bags, backpack pads became sleeping mats.

On Justin's advice, I bought a one-pound waterproof high-thread-count spinnaker-cloth tent, a 6.2-ounce nylon taffeta sleeping bag, a pair of lightweight SPF 40+ hiking pants with mesh-lined thigh vents, and a silicone-coated ripstop nylon backpack.

Since I'd last experienced it, camping—once a way to enjoy the outdoors and commune with nature—had turned into a market for advanced fabrics and technology, all designed to minimize any type of suffering created by interacting with the natural world.

Unfortunately, I would suffer anyway.

# BOOKS THAT KILL

Even after I researched Tom Brown Jr. on the Internet and discovered that nearly every person who ran a respected survival school had once been a student of his—

Even after I read his book *The Tracker,* in which he describes such boyhood adventures as walking miles in a snowstorm wearing nothing but cut-off shorts, surviving weekends in the woods blindfolded, and perching two nights straight on a tree branch avoiding wild dogs—

Even after I learned he'd been shot four times while tracking criminals and found trails leading to some one hundred and sixty dead bodies—

Even after I watched the Tommy Lee Jones movie *The Hunted,* in part based on Brown—

Even as I flew to New Jersey to meet the man . . . —

I had no idea what a force of nature Tom Brown was.

To this day, those two words alone continue to fill me with admiration and intimidation.

After arriving at Brown's Tracker School in the heart of the New Jersey Pine Barrens—the stomping grounds for most of his tales of nearly inhuman feats—I looked around the campground. Scattered throughout were various sweat lodges, wooden shelters, and a shower area consisting of several buckets and mirrors. The buckets were for hauling water from the creek nearby to wash with. The mirrors were for daily tick checks. The woods were crawling with infected deer ticks, which had a nasty habit of jumping onto passing humans, burying their heads in the flesh, gorging on blood, and, if not stopped within two days, leaving a souvenir called Lyme disease.

I found a clearing near an open-air lecture shelter and unpacked my one-pound spinnaker-cloth tent. It was the first tent I'd ever pitched by myself, and at the time, I thought I'd put it together correctly.

By eight P.M. that night, the temperature had dropped to 56 degrees. As we huddled on the benches of the lecture shelter, wrapped in sweatshirts and jackets, Brown stepped up to the podium. Seemingly impervious to the cold, he wore a faded gray tank top with wide cutouts on the sides that ran from his shoulders to his abdomen, exposing as much skin as it covered. Above him, a sign carved out of wood read NO SNIVELING.

"All right—holy shit," Brown bellowed, projecting deep into the woods. "Welcome to Tracker School. This is the center. This is where everything took place."

He gestured grandly to the wilderness around him.

"There's a vast difference between theory and experience," he continued. "If you don't believe me, watch TV. The survivor programs make me vomit. Snivelly assholes!" His face reddened as he spoke. "Between *Survivor, Man vs. Wild,* and *Survivorman,* I've got a nest of snivelers. Surviving in the woods for a year? That's not even getting your feet wet."

In front of him, he had three Poland Spring water bottles lined up. He took a swig from one of them and crushed it dramatically in his hands. "You can walk into any store and there are five survival manuals you can buy that will kill you. Same with those edible plant books. I recover the bodies in their pathetic shelters that don't work because they learned it badly." His unblinking eyes bulged furiously out of their sockets. He almost spit as he spoke his last words: "So forget you have a past."

Unlike Kurt Saxon, Bruce Clayton, and the denizens of the Survivalist Boards, Brown wasn't of the Cold War school of survivalism. He didn't advocate stockpiling dozens of sacks of grain and hundreds of gallons of gasoline and thousands of rounds of ammunition. He advocated taking a stroll in the woods.

"How many people can say that they're free?" he yelled at his students—a mix of marines, hippies, businessmen, and a few psychiatric patients hoping a week of primitive living would restore their balance. "How many people can take a walk for the rest of their lives and never need a damn thing from society?"

Though Brown was fifty-eight, he looked anything but old. He had gray hair parted on the right and plastered by sweat to a face chiseled like an aging superhero's. His flesh was an onion of tan, red, peeling skin in seemingly endless layers, testifying to countless exposures to the elements. His body was thick and powerful, and though it was beginning to atrophy, every slight sag seemed to tell a story few would ever know or experience.

"Why would you want wilderness survival?" His voice raised to a shout. "Because of the what-if question. I'm going to hand you an insurance policy against the what-if question." Now he grew quiet. "By the time I'm done with you, you will be able to survive anyplace with nothing. I'll teach you to build shelters, make fire, forage for food, find water—even if you're in the Sahara."

He was a fantastic performer. One minute he was a drill sergeant, then a preacher, then a crazy old coot. If he delivered even one-tenth of what he promised, attending Tracker School would be the best decision I'd made since deciding to learn self-sufficiency.

Equal parts fact, hyperbole, and self-mythologizing, the Tom

Brown legend is on par with the stories of Paul Bunyan, Geron-
imo, and Mowgli from *The Jungle Book*.

As he tells it, he had a childhood friend named Rick. And
Rick had a grandfather who was a Lipan Apache scout and sha-
man known as Stalking Wolf (usually referred to as Grandfather,
though not to be confused with Grandpa the PT).

At age seven, Brown met Grandfather and spent the next ten
years learning about animals, the wilderness, and Native American life under the elder's almost frustrating figure-it-out-for-
yourself tutelage, known as the coyote style of teaching.

When he was twenty, Brown stripped off his clothes, stepped
into the woods, and lived by his wits for a year. When he returned, he started helping the police track missing children, hik-
ers, and criminals in the forest.

After Brown tracked a suspected robber and rapist who had
eluded some two hundred cops and firemen, the *New York Times*
ran a front-page story on this amazing "27-year-old woodsman"
(though the suspect Brown discovered was reportedly acquitted
of the crimes). A media frenzy ensued, and soon Brown was
doing talk shows, signing a book deal, and opening his own
school.

Antisocial and somewhat misanthropic, Brown used to read
his students excerpts from *Touch the Earth*, a collection of Native American wisdom. But when he noticed he wasn't connecting, he reinvented himself by studying televangelists and the title
character from the campy action movie *Billy Jack*.

"Survival is an art, a philosophy, a doorway to the earth,"
Brown yelled at us. He pulled a handkerchief out of his pocket
and wiped the sweat up his forehead, along the top of his head,
and around the back of his neck. "It's the fucking garden of Eden.
Anyone who says it's a struggle lacks knowledge and skills."

By the time the lecture ended, it was eleven P.M. and the temperature had dropped several more degrees. I walked to my tent and zipped into my ultralight sleeping bag with my clothing still on. With deer ticks around, the last thing I wanted to do was get naked.

The problem with my ultralight sleeping bag was that it was practically useless against the cold. I tried not to snivel as I pulled the nylon top over my head like a cocoon and cinched the drawstring tight. As soon as I generated a little body heat, my bladder began to squeeze urgently. So I extracted myself from my cocoon, stepped into the cold night air, and walked to the outhouse.

All night, the pattern continued: a vague semblance of comfort, followed by a powerful need to urinate. I couldn't understand how anyone enjoyed this hellish torment. Mankind had invented the mattress, the comforter, the asphalt-shingle roof, and climate control for a reason. In comparison with Brown's stories of surviving in the wilderness with no tent, sleeping bag, or even clothes, I felt incredibly lame.

I could order room service like nobody's business, though.

In the morning, I walked to the campfire, where a dozen students had already gathered. I asked them how they slept. "Just fine," they replied.

When I said I'd been cold all night, they looked at me like I was the retarded kid in class.

An older man wearing a military vest with at least eight pockets in the front took pity on me. "You know," he walked up to me and confided, "the secret to staying warm in a sleeping bag is not to wear any clothes."

"Really? I thought clothes would be good for insulation."

"Think about it: your body was designed to warm itself. Your legs, when they're touching, will heat each other."

"But what if you have to go to the bathroom? Isn't it cold getting out of the sleeping bag all the time?"

He smiled and patted the upper left pocket of his jacket. "That's why I carry this." The head of a crushed water bottle poked out of the flap. "Never have to leave the tent."

Though the sight of his emptied urine container revolted me, it also filled me with inspiration. That night, I promised myself, things would be different. I would be prepared.

There was just one last thing that worried me. "What about the ticks?"

The question triggered a group discussion of Lyme disease. According to one of the volunteer employees, half the people working at Tracker School had been infected. "I had it pretty bad," he said. "Half my face was paralyzed for about a year."

My newfound hope drained out of me. I couldn't wait for the week to end. I didn't like camping. I didn't like cold. I didn't like facial paralysis.

I noticed that most of the instructors and volunteers spoke with disdain for the modern world. They called sneakers foot coffins, in comparison with traditional Indian moccasins; they referred to cell phones as ear prisons; and cotton was death cloth, because it loses the ability to insulate when wet. They seemed to believe everyone driving cars and going to offices and watching television in the outside world was unfulfilled because they didn't know these far superior native skills.

One of the lectures that day was on stalking animals. It was taught by Brown's son, Tom Brown III. After describing the areas where we would most likely find game, he told us, "One of the major rules of survival with all living things is conservation of energy."

It was the third time I'd heard that rule at Tracker School. If it was true, then the instructors' scorn for the modern world made no sense. The automobile, the Internet, the nuclear bomb— they're all means of making transportation, communication, and warfare quicker and more efficient. Thus, they all obey the same natural law animals do: conservation of energy.

By the end of the day, I was tired of this primitive superiority complex. Just because a Native American did something two hundred years ago doesn't make it better than what we do today.

When my BlackBerry happened to ring once, even though I didn't answer it, the students and volunteers nearby turned to glare at me in disapproval. Tracker School was turning out to be a cold, inhospitable place, full of pompous hippies and deadly ticks, run by an aging egomaniac. So I decided to flaunt my foot coffins and ear prisons and death cloth in their face. Throughout the rest of the lectures, I sat and texted Katie on the BlackBerry to my heart's delight.

I couldn't wait to get home.

# CAMPING FOR DUMMIES

That night, as if in retaliation for my cynicism, the temperature dropped to 44 degrees and rain began falling in cold sheets. I noticed a student with a crew cut and his shirt off standing halfway outside the lecture shelter. He was smoking a cigarette with a faraway look in his eyes. He seemed completely oblivious to the elements. I envied his tolerance, strength, and centeredness.

I distinctly recall the night that followed as one of the worst of my life. I was sick to my stomach from the three meals of stew they'd served that day and spent half an hour in the outhouse before returning to my tent. I kicked my muddy sneakers off outside and crept into my spinnaker-cloth cell. A few wet spots had already formed on the floor. I quickly inspected the tent but couldn't figure out where the leak was coming from.

Then I took the old man's advice for staying warm and snug. I put an empty water bottle next to my sleeping pad, peeled off my clothes, and quickly slipped naked into my ultralight sleeping bag. I brought the BlackBerry, my only lifeline to the modern world, into bed with me so the rain wouldn't damage it.

Though I pulled the sleeping bag drawstring as tight as possible, I couldn't get warm. The thin layer of down didn't provide enough insulation, and my body couldn't generate enough heat to compensate. When my knees touched each other, they felt like snowballs.

Half an hour later, not only was I colder, but I also had to pee. I reached for the water bottle and felt wetness. When I switched on my flashlight, I saw that my sleeping bag, raised a few inches off the ground by the pad, had become an island floating in a shallow lake. I'd clearly done something wrong when putting my tent up.

I raised myself to my knees, lowered the sleeping bag carefully beneath waist level so it didn't fall into the lake, and unscrewed the cap on the plastic bottle. To make sure I didn't miss, I pressed my dick firmly into the opening of the container and then released. Instantly, droplets of warm liquid began hitting my hands. I couldn't understand why this was happening. Maybe it was performance pressure. I stopped the flow, readjusted, and started again.

This time, it was worse. The urine streamed down the outside of the bottle and into the open sleeping bag below. I stopped again. I didn't know what to do. It was too wet and too cold, and I was too naked, to go outside. Besides, the ticks would be crawling all over me in a second, burrowing their Lyme disease–infected heads into my private parts.

No wonder I'd sworn off camping when I was a child.

I had no choice but to continue. Maybe I'd been pressing the bottle too tightly against myself. I lowered the container a little and emptied the rest of my bladder onto the sides of the bottle, my hands, and my sleeping bag.

It's moments like these that make a man believe in God. Because someone must be laughing somewhere. That night, as I pulled my wet, stinky sleeping bag around me and cinched myself inside, I was a sitcom for the universe.

I fished for the BlackBerry and called Katie to complain.

"I fucking hate it here."

"What's wrong?"

"I'm wet, cold, naked, and covered in urine."

"What happened?"

"My tent's leaking. It sucks."

"You should get a bigger tent, like for six people. With a wood floor."

Maybe I shouldn't have called her. I wasn't in the mood to humor her cheerleader fantasies of what life was like. "Have you ever even been fucking camping before?"

"No. I don't want to get bitten to death by mosquitoes."

"Then let me tell you about the fucking ticks. You'll love them. They suck your blood and give you diseases that make your face freeze."

I proceeded to have a complete breakdown—whining, sniveling, yelling into the phone. I was pissed at Tracker School and every smug moccasin-wearing motherfucker in it. I was pissed at Justin and his lightfuckinguseless gear. I was pissed at nature for creating rain, cold, and ticks. And soon I was pissed at Katie for nothing that was her fault.

By the time I was done with my harangue, the water level had reached the top of the pad and begun dampening the bottom of the sleeping bag, making its thin layer of insulation completely ineffective. On top of everything, I was so worried about Lyme disease that I kept trying to peel off my moles, thinking they were ticks.

Eventually, I lay there, in my 44-degree urine-soaked cocoon, and just gave up. I made the guy from *Into the Wild* seem like Grizzly fucking Adams in comparison.

I was forced to conclude the inevitable: I was a sniveler.

This whole attempt to learn survival was starting to seem like a pipe dream. I was addicted to the three C's: comfort, civilization, and convenience. If the shit hit the fan, I would be the first to get sucked into the blades.

All those stories about Tom Brown living in the woods for weeks at a time at the age of eleven, with nothing but the clothes on his back, seemed impossible.

I rolled onto my side, pulled my knees up to my chest, and heard a splash. I ignored it and tried to go to sleep, despite my bladder's sudden push for more relief. Like a trapped houseguest in a bad vampire movie, I just needed to survive until the sun rose so I could be safe another twelve hours—until night fell again.

## LESSON 46

# WHERE BLACKBERRYS GO TO DIE

As dawn leaked into the tent, I woke up and felt inside the damp sleeping bag for my phone. It wasn't there. I looked over the side of the island, and there was my precious Black-Berry—lying facedown in the lake. That must have been the splash I'd heard the night before. My lifeline to the outside world had been severed. And, as upset and urine-soaked as I was, I had to admit that perhaps justice had been served.

If I was truly going to learn to live without the system, I would have to let go of my dependencies on technology and the outside world. Nature had spoken. The sitcom of the gods had its punch line.

That morning, I asked my usual city-idiot questions around the campfire. "Were you guys cold last night?" "Did your tents leak?" "Were you uncomfortable?" And just like the previous night, everyone replied no. One student said he'd actually found a female volunteer to share his sleeping bag with. I hated him for that.

"I even slept naked to stay warm and that didn't work," I told them.

"That sounds like a good way to get hypothermia," one of the other students replied.

"Are you sure? Some guy told me yesterday that your body keeps itself warm that way."

Another student said that was true only for certain types of sleeping bags. Someone else said it was only true if the alternative was wearing death cloth. And a volunteer informed me that it just plain wasn't true. I didn't bother to ask them about the water bottle.

I guess I'd learned that the wrong survival tip can, in the right situation, kill you.

While we were talking, Brown stepped out of a new black Hummer that just barely fit onto the narrow trail. "Take a look at the ground," he yelled at us. "It'll be the last time you see it that way."

He turned away, then shouted back over his shoulder, "You'll soon be as obsessed as I am."

I wanted the level of comfort with discomfort that Brown and that shirtless smoking guy the night before had. I needed to stop sniveling and toughen up. So what if I was wet? So what if I was cold? So what if I was covered in my own urine? As long as I didn't develop hypothermia, I wasn't going to die from it.

I decided to get advice from the military officers taking the course. Evidently, the government was paying Brown close to a million dollars to design a program specifically for the marines.

The leader of the battalion was a tall blonde in his thirties. After some small talk about what type of enlistee makes a bad marine (people raised by "octopuses" who coddle and smother them) and a good marine (people who follow orders without questioning them), I asked what I really wanted to know: "How can someone learn endurance?"

"I think about that a lot," he replied. "I see some people in certain situations put up with all kinds of pain and humiliation and

struggle, and then in other situations, they crap out after just a little work. True endurance, I think, comes from the inside. It comes from motivation and belief in what you're doing."

I thought about those words and realized that I was resisting the whole camping experience—approaching it with a tight sphincter, as McNeese would say. With my ultralight technology, I was trying to protect myself from nature rather than letting go and trying to get closer to the earth, like Brown preached.

"Do you know what I think when I see a backpacker?" he had shouted at us during our orientation. "I think the same thing when I see a scuba diver or an astronaut. They're aliens to their own environment. If something goes wrong with their equipment, they're dead. They're dead because they don't belong."

Near my campsite, there was a sweat lodge still smoking from recent use. I'd never seen the inside of one before, so I opened the door, crept inside, and sat near a pile of warm rocks in the center. I hugged my knees, rocked back and forth, and tried to empty my head and allow my logic brain to loosen its grip over my wild brain.

Something else Brown had said in one of his lectures popped into my head: "In a survival situation, you've got to know when to let go of your humanness and become an animal. Because if you hold on to thought, there is restriction."

He was right. The more I thought about the cold and the rain and the ticks, the more power I gave them over me. How smart is a duck, really? And an ant—how big can its brain be? If they can figure out how to survive without electricity and running water and gas and fast food, why can't I?

It was only 44 degrees, after all. I could withstand that. And if I took my tent down and asked someone to help me pitch it correctly, I wouldn't have to deal with leaks again.

If I wanted to become a survivalist—in fact, if I wanted to become a human being fully engaged in life—I needed to start enjoying nature instead of fearing it.

In Los Angeles, I didn't let the very real possibility of a high-speed car crash keep me from getting on the freeway. In New York, I didn't let the very real possibility of being mugged keep me from going out at night. So why was I letting a tiny little tick, whose worst effect can be cured with antibiotics, ruin the outdoors for me?

Perhaps I wasn't actually scared of nature, then, but of the unknown. If I wanted to get past this, I needed to stop resisting the one man who promised to make nature known to me.

I left the sweat lodge an hour later with a new attitude—not just toward Tracker School, but toward life. And I didn't even have to sweat for it.

# HOW TO BE 90 PERCENT CERTAIN YOU'LL ESCAPE CANNIBALS

L ater that day, I sat in the lecture shelter—with not just my ear prison gone but my resistance to nature gone—and listened to Brown talk about tracking. He told me that around dawn, according to tracks he'd seen, a doe and a yearling had passed my tent. And I was too busy pitying myself to even notice.

"Watch my feet," he instructed the class. "I'm going to raise my right arm. Now I'm going to touch my nose. I'm going to lower my hand four inches. Now I'm going to breathe."

With each movement, his feet raised or compressed slightly on the ground. And as they did, I realized that footprints convey a lot more information than just shoe size and direction of travel. Every movement people make affects their posture, balance, and weight distribution, which changes the pressure they put on the ground and alters their footprint.

"I will never invade a student's right to privacy." His voice became strident again and his blood pressure seemed to rise, indicating he was switching from teacher to preacher mode. "But some things are red flags. I can pick out breast cancer from a track. I can pick up pregnancy within eleven days—sometimes less. So if I tell you to go to a specialist, please go see a specialist."

Whether his claims were genuine or not, he was truly a freak of nature. The way some people are born to be singers or painters, he was born to be a tracker. Over the course of several lectures, he taught us more than seven hundred different pressure releases made by the interaction between a foot or paw and the earth, how to recognize different animals by their prints, how to track on solid surfaces like rocks and wooden floors, and how to find lost people.

Fact: When lost, individuals generally circle in the direction of their dominant hand. And though they may think they're traveling in a straight line, they're usually circling within the same square-mile area.

After his final tracking lecture, Brown led us into the woods. "I want you to see what's gone on underneath your nose for the last half hour," he yelled at us. "Aliens did not land overnight. Mother Nature does not get goose pimples. If it's a dent in the ground, by God, something made it."

Every few yards, he stopped walking, laid down a popsicle stick, and announced what had happened at that spot earlier— "Here's where a coyote got into a scuffle with a rabbit," "Here's where a shrew peered into a hole," "Here's where a squirrel stopped eating an acorn."

And sure enough, when I got down on my hands and knees and squinted, there was the evidence, glaring at me in prints with the exact pressure releases Brown had just taught us. Some tracks were in dirt, others were on pine needles, and a few were on moss, a log, even a leaf. As Brown had promised, the ground, which was just a monotonous layer of dirt and debris before, was now a library full of stories to be read.

He instructed me to crouch next to the front right paw-print

of a tiny mouselike shrew and study it. It was a small, shallow, shell-like dent in front of a thumb-sized hole in the ground. Once my eyes adjusted to the world of the very small, I was able to see the print in detail—even the thin ridges between each toe and the little black dots where each claw had touched the ground. I could tell where the shrew was heading, how fast it was going, and what it was doing. Eventually, I felt I knew the personality of the animal itself. When I stood up after staring at the ground for a good forty-five minutes, I was able to spot the print even at a distance. It was like learning a new way of seeing.

By this point, the turnaround that had begun in the sweat lodge was complete. I was paying close attention to the lectures, enjoying the outdoors, and looking forward to curling up in my sleeping bag at night—in warm clothes, of course, and minus the plastic bottle. With no phone to run to when I was bored or uncomfortable, I started talking to other students and catching up on survival techniques from the lectures I'd resisted.

As I was standing around the fire one evening, cooking fish that an instructor had taught us how to gut, I found myself immersed in a conversation with the marines.

A younger marine, Luke, was speaking. He had close-cropped black hair, thin lips, and small, sparkling brown eyes. "This is going out on a limb, but I think there will be a revolution in America in the next hundred years."

"Where's it going to come from?" I asked.

"Me," he said without smiling. He paused, then explained. "If you ask anyone in the military, they hate the government. They have all these rules that hold us back and put our lives in danger."

"If we followed the rules of engagement," an enormous older marine named Dave added, "we'd be dead."

Luke told a story about how he'd gotten in trouble with the

government for breaking these rules. His battalion had captured a top-ten Iraqi fugitive, which provoked villagers in the prisoner's hometown to rebel.

"What happened to them?" I asked.

"They ended up looking like that." Luke pointed to the fire, where fish wrapped in smoke-blackened tinfoil were cooking on hot coals.

I could tell he was proud, that he thought it was cool, that it was a Hollywood action movie come to life with him as the star, but all I could think was that those people had futures that were now gone. One of my War Card fears was running into someone as power-drunk and desensitized as Luke, but being on the other side.

Later that afternoon, Brown taught us what he called the sacred order: shelter, water, fire, food. Wilderness survival, he explained, required taking care of each of those needs, in that order of importance.

For the remainder of the week, we were taught how to use nature to fulfill the sacred order, which was the reason I'd signed up for Tracker School in the first place.

Every now and then, usually as the sun set, Brown stepped up to the pulpit and delivered a fire-and-brimstone lecture about the imminent apocalypse. Each successive lecture grew more extreme, until he was telling us, "If the shit hits the fan and you're being hunted by other humans for food, your chance of survival is now ninety percent because you've taken this class."

I wished I could believe him, but I knew I still wasn't a survivor. Not only would I need to intensively practice and internalize the skills he'd taught, but most of them presented a new problem my upbringing hadn't prepared me for. They required the use of a knife.

When I turned thirteen, my aunt bought me a Swiss Army

knife. But my parents immediately confiscated it and said they'd return it to me when I was eighteen. When I asked for the knife five years later, they claimed to have lost it. I think they were hoping I'd have forgotten by then.

Consequently, where other students had no problem whittling functioning traps and fireboards from branches, it took me half a day to carve lumpy, ungainly, barely functioning survival tools. If I wanted the ability to live in the woods with nothing but a knife, I'd need to know how to use one. So I added knife training to my survival to-do list.

By his last lecture, Brown had almost completely transformed from naturalist to mystic, warning that, according to the prophecies of Grandfather, the skies would soon turn red for a week and mankind would be forced to flee civilization to survive. "That's what drives me—fear," he concluded in a dramatic whisper. "Fear that we have run out of time."

Apocalyptic prophecy has been around since the dawn of man—most recently manifesting in an obsession with December 21, 2012, a date on which several different predictions coincide, most notably the end of the Mayan Long Form calendar and thus, claim some of its interpreters, so too the world. Of course, what the sport is really about is not the end of the world, but the end of mankind. And our warnings about it are not examples of our madness, but of our own quest for significance. After all, what could be more meaningful than trying to save the species?

As Brown walked offstage in silence, with tears in his eyes, I looked around the lecture shelter. Almost everyone was wearing a jacket or a hooded sweatshirt. I was still wearing a short-sleeved shirt, and I wasn't shivering or sniveling. I looked for the shirtless smoker I'd seen a few days earlier, but he was nowhere to be found. I asked the marines what had happened, and they told me he'd gone off his medication that week and was taken to a psychiatric hospital because he'd become a threat to other students.

It seemed that the toughest survivors were also the craziest human beings.

That night I slept comfortably, soundly, and warmly. In the morning, when I was called to the podium to receive this course completion certificate, I was actually sorry to leave:

As the van taking me to the airport pulled away from the Tracker School office and made its way to the Garden State Parkway, I gazed through the window at the dense green thicket that ran along the shoulder of the road and realized that, in a single week, my entire reality had changed. I used to think the pavement was my home and the trees and shrubbery off to the side of the road were no-man's-land.

Now, I knew that it could all be my home.

## LESSON 48

# ACTIVITIES FOR IRRITATING YOUR GIRLFRIEND

His voice was cold and gruff. "Are you interested as a dilettante, or do you really want to learn how to use a knife?"

"I want to learn it as a survival skill and as a life skill."

"Good." He seemed strict, with little patience for error. "You're going to thank me when you get two flat tires in the desert and you're trying to figure out how to skewer a rabbit with only a pocket knife, two rubber bands, and some twine."

Once again, curiosity and the power of the odds seemed to have led me to the right person. After scouring the Internet for a knife tutor—only to find cooking classes—I'd visited a store called Valley Martial Arts Supply. The owner, Rafael, had recommended getting in touch with a man in Arizona famous for making the toughest, most effective combat and survival knives in the country. The man was known as Mad Dog.

Mad Dog lived near Prescott, Arizona—the home of not just Gunsite but also the survivalist author Cody Lundin, the shooting instructor Louis Awerbuck, the combat handgun trainer Chuck Taylor, the shotgun manufacturer Hans Vang, and a fringe militia Timothy McVeigh had visited for advice. Depending on who you were, Prescott was either the safest place in the world or the most dangerous.

273

"It's a good idea to bring your gun with you," Mad Dog advised before hanging up, "because shit happens."

While waiting for my week with Mad Dog to begin, I wrote a meticulous schedule, listing a different survival skill from Tracker School to practice each day after I received my knife training. In the meantime, I began finding ways to build my resistance and embody the motto of the Survivalist Boards: "endure—adapt—overcome."

At night, no matter how low the temperature dropped, I never turned on the heat—unless Katie complained. In the daytime, no matter how hot it got, I never turned on the air conditioning—unless Katie complained. When it was dark in the house, I left the lights off as much as possible to develop my night vision—unless Katie complained.

Eventually, we agreed that I would spend as much time as possible in my small backyard, while she lounged in comfort indoors. "Girls just don't like sweating," she'd say whenever I tried to get her to join me outside. "It's sticky and it smells."

So while I worked in the yard, she watched movies like *Saw II* and *The Texas Chainsaw Massacre* on cable TV. I wasn't sure whether her taste in cinema was a subconscious attempt to face her fears or collect new ones.

Though she admirably tried to join me as I slept under the stars one night, her dread of mosquitoes biting her arms, spiders dropping in her mouth, and ants crawling on her skin soon drove her back inside.

I was turning into a nightmare boyfriend. But I was determined to stop sniveling, to toughen up, to become a man. And gradually, without my climate and clothing regulated for maximum comfort at all times, my tolerance began to grow and my skin seemed to thicken.

In the meantime, however, Katie started to get depressed sit-

ting around the house most of the day with no job, no school, and no driver's license. So, in an attempt to get her out of the house and more involved with the obsession that was consuming my life, I invited her to take Mad Dog's knife class with me.

To my surprise, she actually accepted. "He's probably like a violent, mean guy," she said. "But I'll come and learn how to be a Neanderthal with you if you want."

Either she didn't get the importance of having a backup plan, or she understood it better than I did.

While booking her plane ticket, I realized it had been months since I'd heard from Maxwell in St. Kitts, whose bank account I'd perhaps gullibly filled with money. So I e-mailed him to find out if I'd been approved yet.

He replied that he was still waiting to get the title to my apartment from the land registry so he could forward it to the government as evidence that I'd bought the required real estate. So not only did I still lack a passport, but without a title, there was no proof I even owned the apartment.

Concerned that I was being scammed, I called Spencer for advice. But he was in equally bad shape. "I'm getting so frustrated with the whole St. Kitts thing," he sighed. He still hadn't filed for citizenship because the negotiations for the house he wanted had broken down. "You can't get answers from anyone. They're always on holiday. I've been down there so many times, it's become the bane of my existence."

"So what are you going to do now?" If he had a good backup option, I thought I might try to get my money back from Maxwell and join him. But his plan was way out of my league.

"I was thinking, with a bit more money, I could get an island," he said. "If it's at least forty-four acres, I could have it managed by a caretaker, who would basically just be a farmer. Then I'd use half the island for agriculture and meat, and have the bulk stuff ferried out once a month. I did some research, and it's a four-year enterprise to buy it and build it. This way, I'll always have an out."

It must be great to be rich, I thought. When no nation will give you a passport, you can just buy your own country.

# A NINJA'S GUIDE TO BRINGING WEAPONS ON AIRPLANES

*C**ogito ergo armatum sum*—I think, therefore I am armed.
This was Mad Dog's philosophy.

He told us that he liked to go to airport gift shops and see what items he could turn into weapons.

He told us that his daughters wore glass epoxy composite chopsticks in their hair when they flew, so they could sharpen them into daggers if they needed to stab someone.

He told us that he usually traveled with a fighting cane in his hands, which airport security thought was a regular cane.

He told us that anyone could make a killing baton on a plane by wrapping two rolls of duct tape around a tightly rolled magazine.

Listening to Mad Dog, all the FAA regulations prohibiting box cutters and lighters and pool cues and snow globes on planes seemed like a charade, capable of discouraging only amateurs.

Katie sat in the backseat of his truck, petrified. Maybe I'd chosen the wrong survivalist to bring her to.

Mad Dog parked in front of his workshop, MD Labs. Instead of a business sign on the door, there was a notice, as severe as Mad Dog himself, reading WE ARE NOT OPEN TO THE PUBLIC ... IF YOU ATTEMPT ENTRY WITHOUT AN INVITATION OR APPOINTMENT,

YOU WILL BE ARRESTED FOR TRESPASSING IN A FEDERAL CONTRACTOR'S FACILITY.

On the wall inside were pictures of his daughters, all cute girls carrying deadly bladed weapons. The workshop itself was an immense warehouse, the size of an airplane hangar, full of machinery and carbon steel knives in various stages of manufacture.

He lectured us in a slow, measured voice that left no room for things like humor or questions. "A drill bit is two knives in a helix. Scissors are two knives set up in opposition to one another. A saw blade is dozens of tiny little knives. But the finest expression of mankind's most important tool is the fixed blade knife."

"I'm afraid I'm going to hurt myself with a knife," Katie interrupted.

"Why?" he asked. "You have sharp edges all over yourself."

"No, I don't." She was offended by the very thought of it.

"Some of the earliest tools were bones. And look at your nails. I could teach you to defend yourself with those so you could scratch someone's eyes out."

Katie fell silent.

I noticed that Mad Dog never used phrases like "I think" or "I believe" when expressing ideas. To him, they were facts, as clear and present as the 5.4-inch blade he was now pulling out of a slip sheath in his waistband.

It was a knife of his own making, the Bear Cat. "This is like one of my kids or my daughter," he said, deadly serious. "It's not expendable."

He then walked us through the meticulous process he used to make this kid—the band saw that cuts the knife form out of the carbon bar, the milling machine that fine-tunes the band saw cut, the grinder that fine-tunes the milling machine cut.

"This is where the soul of the blade is born," he said as he brought us to his main furnace. Next to it was a long metal box

containing the quench oil that cools and hardens the blades. It was filled mostly with vegetable oil. As for the rest of the liquid, he said, every time he cut himself while working, he dripped the blood into the oil. Thus, every Mad Dog knife was hardened in his own blood.

People like Mad Dog make the best instructors. They are so obsessed with their craft—a table to Mad Dog, for example, is not a table but a hard object shaped by an edged blade—that they don't just give you a lesson, they brainwash you.

More than Spencer, whose strength came from his bank account, Mad Dog was the kind of guy I wanted to be with WTSHTF. A life's savings could disappear overnight, but Mad Dog's strength, conviction, and skills came from within. In a survival economy, they were gold.

I picked up a polished blade nearby. It appeared to be close to completion. I could almost see my reflect—

"If you break that," Mad Dog said coolly, "I'll stick a knife in your throat."

Then again, maybe I'd rather be with Spencer WTSHTF. I looked at Katie to see how she was holding up. "Are you okay?" I asked.

"At first, he seemed kind of crazy," she said, "but when he talks to me, his voice gets softer and he always checks to make sure I'm comfortable. Those are, like, nice-person signals."

We moved to a table in the center of the workshop, where Mad Dog gave us a primer on knives.

He taught us why carbon steel blades hold an edge better than stainless steel.

He taught us about cutting tools, prying tools, and whacking tools.

He taught us about flat grinds, chisel grinds, and convex grinds.

He taught us about the primary bevel, the secondary bevel, the tang, the clip, the ridge line, the ricasso, the choil, the belly, the false edge, and the grind plunge.

He taught us cuts like the chop, push, slice, whittle, tip cope, and edge cope.

He even taught us how to sterilize a knife with sodium hydroxide in order to get rid of DNA evidence.

I never knew a simple blade could be so complex. Like learning tracking with Tom Brown, a language I'd never known existed before was opening up to me. From that day forward, I would never look at a knife the same way again. It was art, science, poetry. It was an entire liberal arts curriculum, but with a practical application.

I looked at Katie's notes. She had begun by taking meticulous dictation, but somewhere between the chisel grind and the primary bevel, her notes had turned to doodles. When Mad Dog started teaching us how to sharpen blades with *Karate Kid*–like repetition, she whispered to me, "Sharpening knives is boring." Then she walked to the bathroom and emerged with an armful of *Maxim* back issues to scour the photos for makeup ideas. I suppose, in her own way, she was working on one of her gender's oldest and most powerful survival skills.

Katie and I woke up early the next morning for a day in the forest learning more advanced knife handling with Mad Dog. As I slipped into my cargo pants, tactical shirt, and gun belt—my entire wardrobe had changed since Gunsite and Tracker School—Katie informed me she wouldn't be joining us.

"What's wrong? You're going to be bored staying in the room all day."

"That's okay." She dropped resolutely onto the bed and pulled the covers up to her chin. "I'm scared of the forest."

After thirteen minutes of pointless debate, I gave in. When Mad Dog picked me up and asked where Katie was, I told him, grumpily, that she wasn't coming because she was scared of the forest.

"She's a survival liability," I sighed.

"Not necessarily. She's an excellent source of protein."

If I was going to be a survivalist, I suppose I'd have to start getting used to these kinds of jokes. That is, if they were jokes.

Mad Dog was wearing dark sunglasses, cargo shorts, and a sleeveless T-shirt that read TO SAVE TIME, LET'S JUST ASSUME I KNOW EVERYTHING.

"Did your family get you that shirt?" I asked.

He nodded, smiled, and began telling me about his background. A softer side of Mad Dog—the side Katie had noticed in his workshop—was starting to come out. Though he had a reputation for being harsh and irascible, especially within the knife community, he wasn't a misanthrope like Tom Brown. He was an artist, compelled to make knives because no one else did it the right way—which was, as his shirt stated, his way.

Born Kevin McClung, Mad Dog grew up in Redwood City, south of San Francisco. "Imagine the mind of a college sophomore trapped in a skinny ten-year-old's body in 1968," he explained. Consequently, he was picked on and beaten up regularly. In fourth grade, when a bigger student threatened to cut him, Mad Dog brought two feet of dog lead chain to school the next day. When the bully confronted him in the schoolyard, Mad Dog ran at him and swung the chain. As he recalled, "The kid's head popped open like a red tomato. He never fucked with me again."

After that experience, Mad Dog always made sure there was something he could use as a weapon in his pocket or within arm's reach, whether it be a steel-jacketed ballpoint pen, a bicycle-lock chain, or a glass soda bottle. In eighth grade, he made his first knife in shop class out of a piece of steel bar stock. At eighteen, he made his next knife by grinding down the edges of a metal file. After working at two veterinarian's offices, two international arms dealers, and a rocket company, he decided to start his own business, Mad Dog Knives.

In the forest that day, he taught me how to use fixed-blade knives, as well as machetes, axes, and saws. After helping me make a spear by lashing a blade to a stick, he showed me how to use the full weight of my body to kill effectively with it.

"You have to learn how to hunt—even if it's just small game— in as many ways as possible, from clubbing carp in shallow lakes to shooting deer with a rifle to killing pigs with a spear to setting traps," Mad Dog explained. "Protein is a commodity. You can trade that for other things you need."

To learn to prepare that protein, I chopped wood and carved some of it into a spit pole and drying rack for meat. Then, using a magnesium fire starter and a knife, we built a cooking fire with the rest of the wood. Finally, we stalked through the forest with a bow and arrow Mad Dog had made, looking for game. Fortunately, we didn't find any.

It was the manliest day of my life. Even the day I lost my virginity didn't feel nearly this masculine.

Afterward, we went back to his workshop, where he gave me a knife exam. He handed me a wooden plank and instructed me to turn it into a spoon. After two hours of laborious sawing, whittling, and gouging, I produced this:

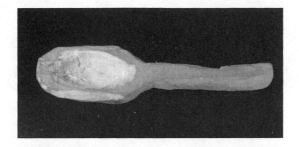

It was a monstrosity. But three days earlier, I could barely even cut the knot off a stick, let alone turn it into a spoon. I felt ready to start practicing the wilderness survival skills I'd learned at Tracker School.

Unfortunately, I still had one more day left with Mad Dog. And it would be a day for which I'd never forgive myself.

## LESSON 50

# THE DOWNSIDE OF THE CIRCLE OF LIFE

"Make sure you push the knife all the way through," Mad Dog ordered. "You don't want it to suffer. If there's a bleeding goat running all over here, I'm going to shoot it"—he patted the .45 on his hip—"and then I'm going to shoot you."

This was horrible. I didn't believe in killing things. Even if I was serving on a jury for the trial of the worst mass murderer in history, I still wouldn't be able to condemn him to die. I don't want anyone's blood on my hands.

Evidently, Mad Dog felt differently.

As he'd promised when I'd first called him, he was going to teach me to slaughter, skin, butcher, and cook animals. And in Mad Dog's world, there was only one way to learn—and it wasn't from books.

I brought Bettie to the hanging tree, straddled her, and removed the leash. She bowed her head and started eating grass and twigs on the ground, unaware this would be her last meal.

"Hold your hand around its mouth and pull the head back to expose the neck," Mad Dog yelled at me.

I was scared shitless. As Tarkovsky said, it's human nature not to do anything irreversible. And there was no going back after this. Of course there are people who hunt for sport regularly, but using a knife was so up close and personal.

I pulled the blade out of its sheath and touched it to the side of Bettie's neck. "Here?" I asked Mad Dog.

Bettie lifted her head from her meal and began to wriggle away. She was beginning to sense something was wrong. "Yes," Mad Dog yelled. "Just do it. Tighten your legs around it. Grip it strongly."

I still had a choice. I didn't have to kill it.

"If you didn't take this goat from the farm, what would have happened to it?" I asked.

"It was raised to be food."

I turned to Katie. I decided to let her be the ultimate moral arbiter of this.

"Is this wrong to do?" I asked her.

"No, because there might come a time where you're hungry and you need to kill something to eat," she replied. I was surprised to hear those words come out of her mouth. She was beginning to understand what this was all about.

I was only fooling myself if I thought that by eating meat almost daily, I wasn't participating in the slaughter of animals. Just as I'd always fooled myself by saying my morals prevented me from voting to sentence anyone to the death penalty. The truth was, I didn't mind other people killing my meat and sending mass murderers to the electric chair. I just didn't want to take responsibility for it myself.

If I wanted to become a survivalist, if I wanted to become a man, if I wanted to make Ayn Rand proud after finally finishing *Atlas Shrugged,* then I'd have to start taking responsibility for myself and my actions. And if I was ever in a survival situation—whether I was hunting deer or forced to live off mice—I would need to know how to do so as efficiently, safely, and painlessly as possible.

In order to honor Bettie in some feeble way, I promised myself I would use every part of her: I'd eat her. I'd wear her. I'd make her bones into tools. I'd name my first daughter after her.

Something was wrong with my head.

"Now," Mad Dog said. "Push your knife all the way through and cut the throat out."

I asked Katie to wait in the car.

My stomach grew light and queasy, my throat tightened with nausea, my head began to spin. I'd come this far. I had to do it. No, I didn't have to do it. I didn't have to do anything. I could just keep it as a pet. But I was going to do it anyway. Why?

Peer pressure.

I squeezed my legs tightly around Bettie's flanks, then grabbed her chin and pulled it up. She wriggled a little but remained docile. I can't recall whether I closed my eyes or not, but the next thing I remember is pushing the knife straight through the side of her neck in one thrust. As it emerged on the other side, something between a scream and a gurgle pierced the air. It was the worst, most horrifying sound I'd heard in my life. I had to end this.

I pushed the knife straight forward, but it got caught on something. Bettie was gurgling more violently now, struggling to break free from the grip of my knees.

"Push through!" Mad Dog yelled. "Push it through!"

I needed to finish the job. If I made Bettie suffer any more than necessary, I'd be even more of an asshole than I already was for doing this. I pushed as hard and straight as I could. I felt the knife tear through something thin and tough, then it emerged from the front of the neck. Blood poured over the knife, over my hand, onto the ground. Bettie's gurgles grew more shrill, more terrifying.

I saw Katie in the truck, her eyes closed and her hands over her ears.

I released my knees and Bettie fell forward. Mad Dog stuck his knife in her throat and sliced deep inside, severing the connection between her head and her spine to make sure she didn't feel any more pain.

Sweat beaded on my body. Saliva burned the back of my throat. The blood on my hands began to turn sticky.

"Why won't that noise stop?" I asked Mad Dog.

"It's just a fluting sound made by air rushing out of the trachea," he explained, always coldly rational.

I looked at Bettie's legs. They were kicking against the ground. I was worried she was still alive. "That's just a nervous system response," Mad Dog said when he saw me staring. "Look at the eyes. It's dead."

"Did I fuck up? Was there something wrong with my cut?"

"Your execution cut on the goat was done as well as it could possibly be done," he said, clapping a bloody hand on my back. "That was a mortal wound. I want to personally congratulate you on a nice clean kill."

But I didn't feel proud in that moment. I felt vile.

Fortunately, Mad Dog kept me too busy to dwell on what I'd just done. That would come later.

As he gave me instructions, I cut Bettie's tendons and firmly lashed them to a board I'd tied to the tree earlier. Using a bottom branch as leverage, I hoisted the board up to a higher branch, until Bettie was dangling upside down from the tree. Then I took my knife and cut a slit from her anus to her solar plexus.

"Get the fascia away, but be careful not to puncture anything and get it all over the meat," Mad Dog instructed, as if the slightest error would ruin the precious protein.

I peeled the skin away from the muscle tissue connecting it to the gut bag, then slowly sliced the bag open, careful not to puncture any organs. Outside of an earthworm in biology class, I'd never looked inside an animal before. It was amazing how well-organized everything was—all these basic shapes hanging around each other in perfect geometry, like a tidy closet. It made me believe in God, who I hoped wasn't watching right now and waiting until I died to punish me for it.

I could just imagine eternity in hell, doomed to getting my throat slit over and over by an angry goat.

Mad Dog talked me through the disemboweling, yelling at me whenever I came close to infecting the meat with goat guts. I found the bladder—a little balloonlike bag half-filled with urine—pinched it closed, and cut it loose. I found the intestines—a dangling strand of sausages—squeezed them, and slid my fingers toward the open end until Bettie's last shit was complete.

One by one, I pulled Bettie's organs out of her body and onto the ground. I cut loose the heart, still warm in my hands as I lifted it out of her chest. Then I severed the lungs, peeled away the liver, and, lastly, removed the esophageal and tracheal tubes near Bettie's neck. This wasn't just a lesson in survival, it was a lesson in biology.

In Bettie's throat, I saw a green, compressed ball, which brought me back to reality. It was the grass and twigs she'd been chewing as she'd died.

I tried not to think about it. All I could do now was make sure I ate and used every part of her.

And named my first daughter after her.

People kill animals every day. Why was I being so sentimental about it? I had to think like Mad Dog: she was a protein source, like a PowerBar with legs.

Katie emerged from the truck and tentatively peeked at the carcass. "You should have gotten a less cute goat," she scolded Mad Dog, and then turned away. I wanted to talk to her. I was worried she'd feel different about me after this.

I made bracelet cuts around Bettie's rear legs, then spent the next ninety minutes carefully skinning her. I wanted to make sure the hide came off in one piece, so I could turn it into a rug or a hat or a miniskirt for Katie.

When I finished, we dropped the skin, with the head attached, into a garbage bag. I now had a skinless, headless, organless goat on my hands. Mad Dog told me where to cut, and I removed the legs, then cut through the spine and removed the rib cage and neck. I brought the remnants of Bettie to the chopping block, cut her rear feet off with an axe, and dropped the last section of meat into a garbage bag.

I was disturbed by how quickly life became meat. This was necessary knowledge, I reminded myself. This was what man did before Safeway and Kroger and Peter Luger Steak House.

We cleaned the area, left Bettie's organs in a pile for coyotes to eat, and climbed back into the truck.

Mad Dog blasted AC/DC and yelled over the music, "Wel come to the circle of life. You're no longer just a bystander or a parasite. You're actively in it. You've demonstrated the ability to kill to feed yourself. The other side of the coin is that you also have to learn how to nurture and grow things to replace what you take out of the environment."

I was glad to hear Mad Dog say that and to know survival wasn't just about chopping down trees and killing animals. "A survivalist is also a conservationist," he continued. "If you just take, you're like a locust."

After dropping the hide off to be fleshed and tanned, we returned to his workshop, started a fire in the backyard, and dropped Bettie's front legs onto the grill.

Finally, with the same knife I'd used to slaughter her, I ate her. In spite of all that trauma, it was the freshest meat I'd ever tasted.

Before I left, Mad Dog wrapped the rest of Bettie in black garbage bags for me to take home, then handed me an axe to keep and a well-worn book called *A Museum of Early American Tools,* which he said would help me build a workshop at home. "You're the last civilian I'm teaching, you know," he said. "After this, I'm just going to do special training for the military."

"Thanks for being so patient with me."

"I thought it would be worse." A smile spread under his mustache. "But with a little more experience, you'll make a fine survivalist."

I was going to miss Mad Dog. In three days, I'd grown a lot under his tutelage. Not only was I now prepared to start the weekly practice schedule I'd made after Tracker School, but I'd probably surpassed Spencer when it came to preparing for hard times. Though I wasn't sure if being more extreme than Spencer was necessarily a good thing.

As I sat on the plane, clutching the wrapped goat meat in my lap, Katie turned to me.

"Are you traumatized?" she asked.

"Strangely, no. Are you?"

"Yeah, a little." She looked away. "It was really cute. And I'm like a Capricorn."

"Then why did you tell me it wasn't wrong to kill it when I asked?"

"I said that because I knew that was what you needed to hear," she said. "But inside, I was thinking, Yes, it's wrong, because it's a cute little thing with a heartbeat." She paused and thought further. "The only reason she let you hold her was because she trusted you. That's fucked up."

Now I was traumatized.

# A WORKOUT PLAN FOR THE END OF THE WORLD

## FIRE SUNDAYS

Before Tracker School, I never knew it was actually possible to make fire by rubbing two sticks together—despite seeing Tom Hanks and an anthropomorphic volleyball do so superbly on *Cast Away*. Now I knew two ways to turn sticks into lighters: the hand drill and the bow drill.

So for my first post–Mad Dog practice session, I drove to the Gabrielino National Recreation Trail, where the Gabrielino Indians once lived off the land, and I gathered branches of cedar and mulefat, along with mugwort leaves for tinder.

I let them dry for a week, then set about creating fire. To make the hand drill, I carved the cedarwood into a rectangular block and cut a small divot on top, near the edge. Then I set it on the ground and fit a long, thin stalk of mulefat over the indentation. I placed the stalk between my palms and rubbed it back and forth so that the spinning created friction and heat against the divot.

Though a lonely adolescence had given me plenty of practice with that rubbing motion, it took two weeks, and many calluses and curse words, before I was able to generate enough smoke to expand the divot into a larger, burnt hole.

After cutting a pie-shaped wedge out of the divot, I was finally ready to start making fire. After two more weeks of tediously spinning the stick, I managed to produce a small ember. I then caught it in a bed of mugwort, cupped it in my hands, and blew until the mugwort burst into flames.

Fire.

I was in awe of this near-miracle of combustion. As long as I had some sort of pithy wood to work with, I would never be without fire again—which meant heat, cooking, and water purification in a survival situation.

If the hand drill is the horse and buggy of firemaking, the bow drill is the Model T—the next step up the evolutionary ladder. To make the bow, I peeled off thin strands of yucca leaf and twined them together. I then tied the yucca cord to either end of a thick curved oak branch.

To make the other three necessary pieces of the drill, I carved a rectangular fireboard out of cedar; a pointed spindle intended to spin in a notch in the fireboard to generate friction; and a small square block to hold on top of the spindle to keep it steady.

Finally, in a somewhat disgusting yet undeniably practical bit of survivalism, I did something I learned in Tracker School. I stuck a finger in my ear, fished out a little wax, and rubbed it on top of the spindle to lessen the friction against the handhold. After all it's a renewable resource.

To generate an ember, I wrapped the string of the bow around the spindle and pulled back and forth to make it spin in the notch of the fireboard. When the yucca cordage I'd spent hours making snapped in seconds, I replaced it with paracord and ended up with this:

The faster speed and more continuous motion generated an ember much quicker than the hand drill did.

Though I was frustrated by the slow progress on my St. Kitts citizenship application (Maxwell was now claiming he had to wait for the prime minister to return from vacation), at least it gave me time to get comfortable with my first skill in the sacred order.

## SHELTER MONDAYS

My initial Monday shelter session lasted from noon until sundown.

To begin the day, I wandered the hills near my house with garbage bags, collecting dead grass, leaves, twigs, and branches. Then I returned to my backyard and stuck two Y-shaped branches in the ground about eighteen inches apart and leaned them toward each other, so that the tops met in the center to form two adjacent Ys.

I nestled one end of a thick, eight-foot-long branch into the Ys and propped the other end up on a log. This would be the ridgepole for my shelter.

I leaned sticks against the ridgepole on both sides so that it began to look like a small tent. Then I covered this ribbing with branches to create a crude latticework. Finally, I dumped all the debris I'd collected on top to end up with this:

Supposedly, three and a half feet of debris will keep a person warm above thirty degrees Fahrenheit. Four and a half feet is good to zero degrees. And six feet of dry, dead debris will keep someone alive at forty below zero—an experience I hope never to have.

I shoved as much remaining debris inside the shelter as I could. Then I crawled in, rolled around to smash it flat, and filled the space with debris again. After repeating this process a few more times, I built a small door frame out of leftover sticks and leaves, then used one of the debris-filled garbage bags to seal the opening. According to Brown, a good way to keep warm in an emergency is to stuff your clothing with fistfuls of debris.

Fact: some homeless people do the same thing, but with crumpled pages of newspaper.

That night, I put on long underwear, jeans, and a hooded sweatshirt and waited patiently as Katie tried to stop me. "There

are microscopic bugs living in the leaves. You're basically sleeping on a bed of bugs. Don't do it. It's disgusting."

"I'll fumigate it with a smoldering branch."

"You should wear a surgeon's mask too. So the bugs and spiders don't go into your nose and mouth." She noticed I was laughing, then thrust her lower lip forward. "You think I'm kidding, but I'm not. I'm sure whatever they're eating isn't as good as your blood."

After promising to come back inside for a dust mask if I needed one, I crawled into the debris hut. I scrunched into the leaves, pulled the garbage-bag door shut, and prepared for bed.

After an hour, I was cold, itchy, and fantasizing about my warm, clean, indoor bed with a girlfriend in it. But instead of sniveling, I put on shoes, grabbed a flashlight and a garbage bag, and searched for more debris to insulate the hut. I'd gathered most of the dead leaves near my house already, so I filled my bag with yard trimmings from a neighbor's recycling bin, dumped it over gaps in the hut, and crawled back inside, exhausted.

But before I could fall asleep, I heard something scratching outside. The more I tried to ignore it, the louder it seemed to get, until I was wide awake, worried that at any moment a rat would burst through the debris and scamper across my face. Maybe I needed that surgeon's mask after all.

If I'd been in a real survival situation, I suppose I would have been excited by the prospect of breakfast in bed. But since I wasn't, I rustled the debris and shook the hut. But the scratching didn't stop. All I succeeded in doing was shaking loose some of my precious insulation. The city had its own natural world, and in that moment it seemed much dirtier and more aggressive than the natural world of the forest.

Eventually, I drifted into a deep sleep that lasted until sunrise.

It was amazing to think I'd just spent so much money on a house in St. Kitts yet was able to build this little studio apartment for nothing. All I needed now was an exterminator.

Fortunately, trapping small game was on Friday's agenda.

## WATER TUESDAYS

Before Tracker School, I never knew that cutting a wild grape-vine would produce water. Or that chewing thistle stalk (after removing the thorns, of course) would temporarily mitigate thirst. Or that the small amount of liquid in cactuses is generally too bitter to drink. Or that if I were ever stranded at sea, I could drink fish spinal fluid, the liquid around fish eyes, and turtle blood. Or that I shouldn't eat meat when dehydrated, because breaking down food leads to further water loss. Or that sucking on a pebble with my mouth closed would conserve water and lessen the feeling of thirst.

Most of all, I didn't know that it was possible to make water out of basically nothing. So for my first liquid Tuesday, I built a solar still.

In the morning, I dug a hole roughly two-feet deep and placed a cup in the middle. Then I grabbed green leaves and grass, and scattered the vegetation in the hole around the cup. Supposedly, if there's no foliage, the solar still will still work if you pee in the hole. And if stranded at sea, you can place the cup in a larger container partially filled with salt water.

I covered the hole with a clear garbage bag, secured the plastic in place by burying the edges in dirt, and placed a stone in the center so that the garbage bag dipped inward over the cup.

Incredibly, when I checked the still that night, water had condensed on the inside of the plastic wrap, dripped into the cup, and produced this:

The following Tuesday, I attempted a similar process. But this time, I wrapped a clear garbage bag around a leafy tree branch. Two days later, the bag contained much cleaner, better-tasting water than the solar still had produced, though only half the amount.

I thought of all the movies I'd seen with parched men crawling through the desert, begging for water. If only they'd known how to make a solar still, then things would have been different. They'd have been begging for garbage bags instead.

## THE WEDNESDAY SURVIVALIST
## DINNER PARTY

On Wednesdays, to make my catastrophe training a little less lonely, I invited friends over for a dinner party. Each week, I attempted a different method of primitive cooking as Katie fretted in the background, worried I'd catch the shrubbery, the house, or one of her hair extensions on fire.

Fortunately, thanks to my CERT training, I had a class-A extinguisher next to my fire pit, and I knew how to use it. I'd also bought four containers of Thermo-Gel. Supposedly, if I sprayed the chemical on the outside of my house during a wildfire or a cooking party gone wrong, it would be safe from the flames.

For my first dinner, I tried a rock boil, which consisted of heating rocks in the fire, then dropping them in a pot of water to make goat stew. I decided not to tell the guests where the goat came from.

The following week I tried coal cooking, in which I raked a bed of coals away from the flames and simply dropped raw steak on top.

Next I tried a rock grill, which was an improvement on the above procedure, because cooking steak on top of hot rocks instead of coals resulted in less ash on the meat.

Then I tried spit cooking, in which I thrust two Y sticks in the ground on either side of the fire, and impaled a chicken on a third stick that rested on top. For the first time, I noticed neighbors observing me from their third-floor window. I hoped they wouldn't call the fire department. I was most likely violating several, if not all, of the Los Angeles fire codes.

The following week I tried a steam pit. I dug a hole in the ground, lined it with rocks, and built a fire on top. When the fire burned down to coals, I covered them with six inches of grass and leaves. Then I lowered a chicken on top, added another layer of vegetation, and sealed everything airtight with a wooden board, a tarp, and a thick layer of dirt.

For this particular dinner party, I left a note in my neighbors' mailbox inviting them. I figured it was the polite thing to do, given that I'd dug up an area that may have been their lawn to make the pit.

And, finally, I tried a clay bake, in which I encased a stuffed

chicken in an airtight dome of clay, built a log-cabin fire around it, and let it cook for ninety minutes before cracking it open.

Tried is the key word here. Because initially, every one of these simple cooking methods was a complete disaster. Perhaps this shouldn't have been a surprise, since I can't even make popcorn in the microwave without burning it.

The meal typically began at nine P.M. with hungry guests receiving whatever edible scraps I could gather from the undercooked wreckage of my cooking experiment while Katie called Domino's to order pizza.

But gradually, after enough of my irritated friends pitched in to help, I became more comfortable around the fire than I'd ever been in the kitchen. The result was some of the most tender, flavorful meals I'd cooked in my life.

## TRACKING THURSDAYS

At Brown's suggestion, I built a tracking box in my backyard. It was basically a four-by-eight-foot sandbox. On Thursdays, I made prints in the sand and studied the tracks.

Sometimes I ran. Other times I moved in a slow stalk I'd learned at Tracker School. And after a few weeks, I'd clap my hands or bend down to pick up keys to see how it affected the print.

Though it seems unlikely to be able to read that much from a footprint, the impressions in the sandbox were deep and detailed. In a print like the one below, for example, the amount of sand thrown off the front of the foot tells you that I came to a sudden stop and the cracks to the left of the track are a sign that I then turned right:

Though the goal of these weekly practice sessions was to embody Kurt Saxon's definition of a survivalist—"a self-reliant person who trusts himself and his abilities more than he trusts the establishment"—I began to notice a related and unexpected side effect. I was developing the sense of centeredness I'd noticed in some of the students at Tracker School. When I went out with friends, I didn't need to dress up. I didn't need to be at the coolest club in town. I didn't need to drink. I didn't need to talk to anyone. Wherever I went, I brought myself, and that was enough.

I was turning into either Clint Eastwood in *The Good, the Bad, and the Ugly* or the Unabomber. I wasn't sure which yet. Especially after I started hunting game in my backyard.

## PRIMITIVE HUNTING FRIDAYS

A British naturalist I'd met at Tracker School named Lee had told me that, in his opinion, I needed to master three essential qualities and three essential skills for wilderness survival.

The three qualities were nature awareness, physical fitness, and self-mastery. The three skills were proficiency in the hand drill, the debris hut, and the throwing stick. Since I'd already

been working on the first two skills, mastering the throwing stick became Friday's goal.

The steps to making and using a throwing stick are as follows:

1. Find a stick.
2. Throw it.

In a survival situation, bringing down a deer isn't necessarily a good thing because it's too much meat for one person. As Tom III put it, "The best refrigerator is on the hoof." So the throwing stick is for nailing rabbits, raccoons, squirrels, and other small meals.

The sidearm throw is a Zen art and should be done by instinct as soon as the animal comes into view. Hesitating for even a fraction of a second usually means missing by a wide margin.

On my first Friday, I found the ideal stick on the side of the road near my house. It was heavy, straight, wrist-thick, and about two and a half feet long. That night, I drove to a parking lot nearby and practiced throwing it sidearm and overhead at different targets. Eventually, I could strike almost any stationary object from twenty yards away, though when I saw my first moving target—a raccoon—I didn't have the heart to brain it. Especially since, if I did, I'd have to eat it—or serve it to my friends at the next dinner party.

I was truly regressing as a human being.

Other Fridays, I made snares and deadfall traps and planted them in my yard, hoping to catch the rat that had disturbed my sleep in the debris hut. Though the contraptions were inventive—the rolling snare, for example, was designed to send victims hurtling into the air on the end of a sapling—I was never able to catch the rodent or any of its friends, despite adding bait like cheese, peanut butter, and Bettie.

# SATURDAY EDIBLE PLANT WALKS

Originally, I tried to forage for food by myself. I planned to choose an edible plant to learn about each week, then set off in search of it with a Peterson Field Guide tucked under my arm. But on my first hike, when I was unable to find more than one dandelion—I'd hoped to gather hundreds to make dandelion wine in my bathroom—I decided to seek outside help.

A Google search led me to the School of Self-Reliance run by Christopher Nyerges, who organized weekly edible plant walks in parks and forests around Los Angeles.

I met him and ten other students the following Saturday afternoon at the Arroyo Seco trailhead in Altadena. He wore a blue button-down over a white polo shirt and a straw brimmed hat with a purple bandana wrapped around it. Over one shoulder was a backpack with the tip of a bow drill sticking out. He looked equal parts golfer, zoologist, and Indiana Jones.

Nyerges led us in a mile-long loop, stopping every now and then to feed us a flower, leaf, or stem. He was an unlikely guide, because he didn't appear to enjoy teaching and seemed irritated by questions. He did, however, enjoy having an audience to impress.

We ate the little yellow flowers of mustard plants, which were tasteless except for a slight kick on the way down; the sweet, juicy celerylike buds that lay inside the stalk of young bull thistle; the succulent, cucumber-like roots cut from stalks of cattail; and the delicious, jicamalike fruit of the yucca flower. It seemed like at least a third of the plants we passed were edible.

Then Nyerges stopped in front of what looked like anorexic parsley. I'd seen the plant often that afternoon and wondered if it was edible.

"What's this?" he asked.

Mark Forti, a filmmaker and a regular on the walks, responded, "California parsley surprise."

"If you eat this," Nyerges said flatly, "you'll be dead in half an hour. It's hemlock."

It was amazing how ubiquitous the plant was. If you wanted to kill someone, all you'd have to do is take them on an edible plant walk, then feed them a stalk of "parsley." Some fifteen minutes later their limbs would go numb, followed by unconsciousness and death.

Instead of showing us how to recognize each plant and discussing their many culinary and medicinal uses, Nyerges enjoyed lording over us with his knowledge.

"See that lily?" he'd say. "Pretty, right?"

After we all agreed, he'd rub it in our faces: "If you eat that lily, you die. Most lilies will make your throat swell and suffocate you."

A few minutes later, he was in front of poison oak. "Just standing downwind of this will get you infected," he cautioned. "But I can eat it."

Then he ripped off a leaf in his bare hands and put it in his mouth. In that moment, he joined the ranks of Tom Brown and Mad Dog in my pantheon of survivalist superheroes.

"How did you do that?" I asked, incredulous.

"After a while, you develop an immunity to it."

A few steps later, he pointed out a raggedy green plant that stood inches off the ground. "You could have a party with that." The plant, he went on to explain, was jimson weed, which can be made into a tea that causes psychedelic hallucinations (though too large a dose is fatal).

Every plant we encountered, it seemed, could nourish us, kill us, or get us high. It was like mother nature's own ghetto.

After two months of survival cross-training and walks with

Nyerges, not only did I finally feel proficient with all four elements of the sacred order—shelter, water, fire, food—but I felt more healthy, alive, and comfortable in my skin than I ever had. I'd spent most of my life avoiding long walks, manual labor, and kneeling in the dirt. If it weren't for this training, I never would have realized that these were exactly the things my body was designed to do.

# HOW A CAN OF BEER WILL KEEP YOU WARM

On future walks, Nyerges taught us how to make archery bows from willow branches, rub yucca leaves under water to produce a frothy green soap, identify edible seaweed on the beach, and make fire with soda and beer cans. For this latter feat, he turned an empty Coke can upside down, polished the surface, held tinder on a stick just above the concave bottom, and angled the reflective aluminum into the sun until it lit the tinder.

Even if I never had to use any of these skills in my lifetime, at least they'd make me a more competent dad—and my kids wouldn't be peeing all over themselves in leaky tents, crying to their girlfriends about how they want to go home.

While talking with Forti, the filmmaker, on one of the walks, I asked him what kinds of movies he made. I wasn't expecting much. At best, I figured maybe he'd directed something featuring a talking animal that excelled at a human sport. But it turned out that he was working on one of the most fascinating documentaries I'd ever heard of.

Forti said he'd spent the last seven years traveling around the world to see how different cultures responded to what he believed were the five major questions of existence:

1. What concerns you the most in life?
2. What does it mean to live a good, fulfilled life?
3. What's the problem with mankind and what's the solution? (Interestingly, Forti added, out of nearly a thousand interviews in over a hundred different cultures, no one replied that there was no problem with mankind.)
4. What happens after you die?
5. If there's a God, what's he like?

"I'm curious," I asked after he explained each question. "When you talk to people who've been the target of a genocidal campaign or experienced the extremes of man's inhumanity to man, do they still have faith in God and goodness?"

"I've spoken to people in Rwanda who survived the genocide. And I've spoken to people who've survived acts of God, like in Sri Lanka after the tsunami. And I've found that suffering usually draws people closer to God and gives them more faith. I think that the main driver in the human spirit is hope. Man can endure anything if he has hope."

I was reminded of *Man's Search for Meaning,* the book Viktor Frankl wrote about surviving Nazi concentration camps, and how he said that the most important survival skill to have was faith. As he put it, "Woe to him who saw no more sense in his life, no aim, no purpose, and therefore no point in carrying on."

I wondered, as we walked on, what my aim in life was and why I was so determined to survive. What was it all for?

The answer came to me instantly: I wanted to survive because I wasn't done living.

If our life span is a movie that begins with a tiny screaming neonate and ends with a shriveled, arthritic geriatric, then I don't want to leave in the middle. There's romance, horror, adventure, comedy, fantasy, family, and, most exciting of all, suspense still

to come. And I want to see it all, until the very last credit rolls and the screen goes dark.

A lot of people are driven by the belief that they're special, that they matter, that there is a reason why they're here. And they're sustained by their motivation to have this belief affirmed, which is why when someone challenges it, they want to take them down, whether through gossip, ostracism, aggression, or terrorism.

To me, that sounds like a lot of work and not much fun. All I believe is what I know for certain: that I'm alive, so I might as well make the most of it. And I'm driven by the simple fact that I hate leaving things unfinished.

Unfortunately, that's exactly where I was on my survival quest. Everything was incomplete, and I was stressed. It had been over a year since I'd applied for my St. Kitts citizenship, and whenever I e-mailed Maxwell, his only response was "Be patient."

But it was no longer possible to be patient: every morning I read the news, I saw more evidence of dark times coming. Everywhere I went, I heard of more billionaires like venture capitalist John Doerr predicting and preparing for an apocalypse. Every time I talked to Spencer and Mad Dog and the regulars on the Survivalist Boards, they found new flaws in my plans, holes in my training, contingencies I'd overlooked.

At times, it seemed there would be no end to the amount of things I needed to do to prepare. Learning to survive evidently meant learning every essential skill mankind had developed on its journey from *Homo habilis* to civilized humanity.

And I wasn't opposed to doing that.

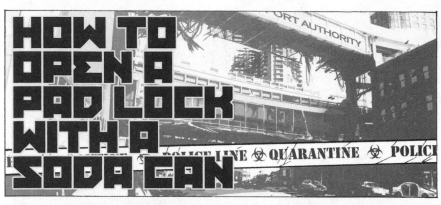

# HOW TO OPEN A PAD LOCK WITH A SODA CAN

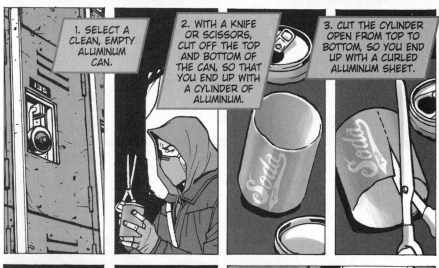

1. SELECT A CLEAN, EMPTY ALUMINUM CAN.

2. WITH A KNIFE OR SCISSORS, CUT OFF THE TOP AND BOTTOM OF THE CAN, SO THAT YOU END UP WITH A CYLINDER OF ALUMINUM.

3. CUT THE CYLINDER OPEN FROM TOP TO BOTTOM, SO YOU END UP WITH A CURLED ALUMINUM SHEET.

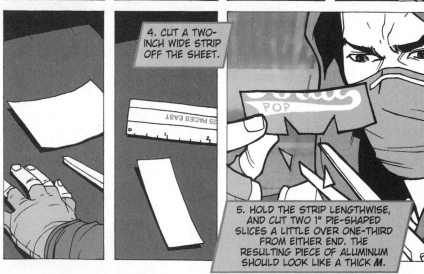

4. CUT A TWO-INCH WIDE STRIP OFF THE SHEET.

5. HOLD THE STRIP LENGTHWISE, AND CUT TWO 1" PIE-SHAPED SLICES A LITTLE OVER ONE-THIRD FROM EITHER END. THE RESULTING PIECE OF ALUMINUM SHOULD LOOK LIKE A THICK *M*.

6. FOLD THE UNCUT PART OF THE SHIM IN HALF. THEN BEND THE TWO END TABS ON EITHER END OF THE SHIM UPWARD, SO THAT YOU END UP WITH A LONG PIECE OF ALUMINUM WITH A SINGLE, ROUNDED TAB PROTRUDING FROM THE CENTER.

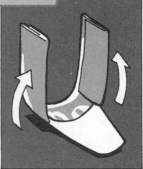

7. YOU NOW HAVE A PADLOCK SHIM. BEND THE SHIM IN HALF, TAB SIDE DOWN, AROUND THE SHACKLE OF A LOCK. MAKE SURE IT'S CONTOURED CLOSELY TO THE CURVE OF THE SHACKLE.

8. PUSH THE TAB DOWN INTO THE CREVICE BETWEEN THE SHACKLE AND LOCK.

9. ROLL THE SHIM AROUND THE SHACKLE UNTIL YOU FEEL THAT IT'S OVER THE CLASP, THEN PUSH GENTLY DOWN AND THE LOCK WILL POP OPEN. IF THIS DOESN'T WORK, TRY THE OTHER SIDE OF THE SHACKLE. A SMALL FRACTION OF LOCKS MAY REQUIRE A SHIM ON EACH SHACKLE.

*TO BE CONTINUED...*

LESSON 53

# SURVIVAL USES FOR MUSIC CRITICS

My house had completely transformed.

My yard was no longer a place to get a suntan. Instead it was filled with a fire pit, a tracking box, a debris hut, shovels, stacks of wood, piles of debris, and newly planted vegetables and fruit-tree saplings.

My basement was no longer just for storage. With Mad Dog's *Museum of Early American Tools* book as my guide, I'd started building a rudimentary workshop.

My closet was no longer for fashion. Cargo pants, wool hunting jackets, ghillie suits, and tactical belts swung from the hangers. The floor was littered with boxes containing flashlights, rechargeable batteries, bullets, gun-cleaning kits, night-vision goggles, and half a dozen folding and fixed-blade knives, along with sharpening rods and stones.

In my garage, there was barely enough room for my car between the Rokon and the boxes of emergency supplies. And the rear seat of my Durango was no longer a place for passengers but a shelf for my bug-out bag, first aid kit, and CERT uniform.

Even the walls of my house were no longer intact. With my early American tools I'd built a hidden panel, behind which I'd stashed my pistol, rifle, shotgun, and metal-detector-proof knives. In the backyard I created a trapdoor with a hidden crawl-space, which was big enough to hide not just supplies but, if necessary, me.

I'd clearly gone off the deep end, and I had no intention of stopping. Not when I was so close to slaying the demon of Just in Case.

Just in case of nuclear attack, I'd bought potassium iodate tablets online to lessen the effects of radiation poisoning.

Just in case of chemical attack, I'd gotten a friend on the Survivalist Boards to send me DuoDote injection pens to stop the symptoms.

Just in case of an armed attack, especially while trying to escape on the Rokon, I used a connection of Mad Dog's to purchase an Enforcer concealable bulletproof vest from U.S. Armor.

Just in case Katie or I became ill WTSHTF, I'd driven to Mexico to stockpile antibiotics (though some survivalists prefer to get them from feed stores instead).

Just in case I needed to evacuate from L.A. by water, I'd contacted the American Sailing Association to take their Basic Keelboat Sailing 101 class.

Just in case of Just in Case, I'd learned to tie basic knots, grind grain, bake bread, catch fish, sew clothing, smoke meat, make preserves, and can food.

All my life, I'd never had to do anything practical. If something in my house or apartment didn't work, I called a repairman or landlord. If my car broke down, I called AAA. If I was hungry, I had food delivered. If I needed something affordable, I bought it online. If it wasn't affordable, I used credit. My tools were the telephone and the Internet, which instantly summoned the services of other people.

But as the world of survivalism opened up, I began to realize that I'd been rendered completely helpless by convenience. Maybe the instructors at Tracker School had a point. Life was more exciting now that I was learning to handle just about anything that came my way without having to depend on anyone else. Two decades after puberty, I was finally becoming a man.

As my obsession with figuring out how to do everything myself intensified, I drove to a self-sufficient community called the Commonweal Garden outside San Francisco that was part of what's called the permaculture movement. I wanted to learn how to design a completely sustainable life from scratch. On the roof of one of the houses there, for example, there was a rainwater catchment, which fed water into a shower below. The runoff from the shower filled a pond, which supported ducks. The ducks ate bugs off strawberries in the garden, which were served for breakfast. The leftover breakfast scraps were dropped into a bin of worms, which were used to feed fish that maintained the balance in the pond. And the worms' waste was used to fertilize the strawberries.

It was a perfect, closed, interdependent system, a microcosm of planet Earth. I stuck around for three days, trying to learn all I could from the people who ran it. And I discovered, much to my distress, that if I wanted to be a more long-term-minded survivalist, rather than killing Bettie for meat, I should have learned to breed and milk her.

"It's bad that you killed a girl goat, because she could have had

babies," Katie had told me at the time. "Guy goats can't do that. There are guy goats with sperm everywhere."

So as soon as I returned home, I went to the website goat-finder.com and began researching the possibility of raising a female goat, much to Katie's delight. Good thing I didn't kill a whale.

In college, I'd seen a documentary called *The Emperor's Naked Army Marches On,* about a Japanese director trying to find out what happened to two soldiers from his regiment who disappeared while serving in New Guinea during World War II. Eventually he tracks down their commanding officers, and after much badgering they admit that the platoon was trapped without food, and at least one of the missing soldiers was killed and eaten so the higher-ranking officers could stay alive.

Watching the film triggered an old fear I used to have as a child, ever since being forced to watch movies like *The Day After* in school and learning about historical aberrations like the Donner Party. I wondered: If I were part of a group that was stuck somewhere with no food, would I be the eater or the eaten?

And I was forced to admit every time, sadly, that I would be one of the eaten. As a student, and then as a writer, I possessed no skills that would be useful to a starving group. Unless, of course, they needed a music critic.

But now everything had changed. My training was making me eligible to lead that group. And as leader I would decree that nobody get eaten—because there's enough lamb's-quarter and cattail and dandelion out there to keep us all alive. And if there isn't—because, say, the earth is scorched—well, if there's a tasty-looking music critic in the group, I can show them exactly how to butcher, skin, and gut him.

What was wrong with me? I was turning into Mad Dog.

I thought of all the postapocalyptic movies I'd seen: *The Omega Man, Things to Come, Mad Max, 28 Days Later, Night of the Comet, The Matrix, Twelve Monkeys, Waterworld, The Day After Tomorrow, The Last Man on Earth, Day of the Dead, A Boy and His Dog, I Am Legend, Doomsday,* and on and on and on. And what did almost every one of them have in common? They took place primarily in a city. And the few that didn't were set in a desert wasteland, with only a little shrubbery to gnaw on—or, in one notorious case, in a water wasteland. There wasn't a lamb's-quarter or cattail plant in sight, let alone enough debris to make a shelter or sticks to make a hand drill or natural resources to sustain a permaculture community.

I may have thought I was ready to survive World War III, but there was one major flaw in my thinking. People like Tom Brown, the primitivists, the naturalists, and the permaculturists lived by certain rules. One of those rules was that anything that grew in nature was good and anything made by modern man was bad. But that wasn't reality.

In a real-life SHTF scenario, it will be hard to avoid the detritus of society. A sharpened bicycle spoke will probably make a better weapon than a stick. A length of electrical wire will make stronger cordage than a yucca leaf. And foam padding ripped from the seats of an abandoned truck will provide warmer insulation than debris. Wilderness survival was good for the forest. But what about the urban jungle?

# SECRETS OF
# ESCAPED FELONS

Kelly Alwood didn't say a word as he handcuffed my hands behind my back, opened the trunk of a rental car, and ordered me to get inside. With his shaven head, which looked like it could break bottles; his glassy green eyes, which revealed no emotion whatsoever; and the .32 caliber pistol hanging from a chain around his neck, he didn't seem like the kind of person to cross.

As he shut the trunk over my head, the blue sky of Oklahoma City disappeared, replaced by claustrophobic darkness and new-car smell. Instantly, panic set in.

I took a deep breath and tried to remember what I'd learned. I curled my right leg as far up my body as it would go and dipped my cuffed hands down until I could reach my sock. Inside, I'd stashed the straight half of a bobby pin, which I'd modified by making a perpendicular bend a quarter inch from the top. I removed the pin, stuck the bent end into the inner edge of the handcuff keyhole, and twisted the bobby pin down against the lever inside until I felt it give way.

As I twisted my wrist against the metal, I heard a fast series of clicks, the sound of freedom as the two ends of the cuff disengaged. I released my hand, then made a discovery few people who haven't been stuffed inside a trunk know: most new cars have a release handle on the inside of the boot that, conveniently, glows in the dark. I pulled on the handle and emerged into the light.

"Thirty-nine seconds," Alwood said as I climbed out of the trunk. "Not bad."

I couldn't believe classes like this even existed. In the last forty-eight hours, I'd learned to hot-wire a car, pick locks, conceal my identity, and escape from handcuffs, flexi-cuffs, duct tape, rope, and nearly every other type of restraint.

The course was Urban Escape and Evasion, which offered the type of instruction I'd been looking for to balance my wilderness knowledge. The objective of the class was to learn to survive in a city as a fugitive. Most of the students were soldiers and contractors who'd either been in Iraq or were about to go, and wanted to know how to safely get back to the Green Zone if trapped behind enemy lines.

The class was run by a company called onPoint Tactical. Like most survival schools, its roots led straight to Tom Brown. Its founder, Kevin Reeve, had been the director of Tracker School for seven years before setting off on his own to train navy SEALs, Special Forces units, SWAT teams, parajumpers, marines, snipers, and even SERE instructors. As a bounty hunter, his partner, Alwood, had worked with the FBI and Secret Service to help capture criminals on the Most Wanted list.

For our next exercise, we walked inside to a shooting range behind the classroom where an obstacle course had been set up. Alwood handcuffed me again, adding leg chains to my feet. I then ran as fast as I could through the course, ducking under and climbing over chairs and benches, simulating a prison escape. By shortening my stride and putting extra spring into each step, I found it easy to sprint across the room. I knew those childhood jump-the-bum games would come in handy.

"You look like you've done this before," Reeve joked.

Though I was hopelessly out of my element when it came to wilderness survival, I actually wasn't too incompetent in this class. Because I'd lived in cities all my life, I had at least some semblance of street smarts.

"We're nine meals away from chaos in this country," Reeve lectured afterward, explaining that after just three days without food, people would be rioting in the streets. "With gas and corn prices so high, the events of the last six months have made it much more likely that you'll be needing urban escape and evasion skills in this lifetime."

To prove his point, Reeve told us of gangs of armed looters that ransacked neighborhood after neighborhood in New Orleans during Katrina. "One of the police officers there told me they shot on sight three people out past curfew," he added.

For some reason, I was more disturbed by the idea of killer cops than marauding gangs. Maybe it was because of the recurring nightmares I used to have as a teenager about being mistaken for someone else and taken to jail. In the dreams, I'd be so petrified during the ride to prison that I usually woke up in a cold sweat before I ever made it there. Since then, I'd come to realize it wasn't actually jail I was scared of in those dreams, but the loss of freedom that it represented.

As the sun set, we drove to an abandoned junkyard, where Reeve let us practice throwing chips of ceramic insulation from spark plugs to shatter car windows, using generic keys known as jigglers to open automobile doors, and starting cars by sticking a screwdriver in the ignition switch and turning it with a wrench.

As I popped open the trunk on a Dodge with my new set of jigglers, I thought, This is the coolest class I've ever taken in my life. If I'd had these skills in school, I would have been playing much more fun games than jump-the-bum, like hotwire-the-teacher's-car or break-out-of-the-juvenile-detention-home.

Over a barbecue dinner later that night, Reeve asked why I'd signed up for the course. "I think things have changed for my generation," I told him. "We were born with a silver spoon in our mouths, but now it's being removed. And most of us never learned how to take care of ourselves. So I've spent the last two years trying to get the skills and documents I need to prepare for an uncertain future."

I'd never actually verbalized it before. I'd just been reacting and scrambling as the pressure ratcheted up around me. Reeve looked at Alwood silently as I spoke. For a moment, I worried that I'd been too candid. Then he smiled broadly. "You're talking to the right people. That's what we've been thinking. Kelly has caches all over the country—and in Europe."

On the first day of class, Reeve had taught us about caches—hiding places where food, equipment, and other survival supplies can be stored away from home, whether buried in the ground or stashed in a bus-station locker.

"The thing with caches is that you have to be able to survive if one is compromised," Alwood explained. "So each one has to contain everything you need: gun, ammo, food, water."

"You'll need lots of ammo," Reeve added, "because that will be the currency of the future."

I pulled out my survival to-do list and added, "Make caches."

I'd noticed that the way people prepare for TEOTWAWKI has a lot to do with their view of human nature. If you're a Fliesian like Alwood and Reeve and you think that without the rules of society to restrain them, people will become violent and selfish, then you'll build a secret retreat, stockpile guns, and start a militia. If you're a humanist like the permaculturists and believe peo-

ple are essentially compassionate, then you'll create a commune, invite everyone, and try to work in harmony together.

Just as I had my triangle of life and Tom Brown had his sacred order, Alwood and Reeve had their own list of essentials: water, food, defense, energy, retreat, medical, and network.

By this point, I already had at least the basics of most items on their list covered—with one glaring exception. I didn't have a network.

I'd found no groups where I felt like I belonged. The billionaires were out of my league. The PTs were too paranoid about Big Brother. The survivalists were too extreme about guns and politics. The primitivists were too opposed to technology and modern culture. And the growing tide of 2012 doomsdayers seemed more interested in trying to prove it than doing anything about it.

And unless you're Robert Neville in *I Am Legend*—and even he died at the end—the best way to survive WTSHTF will be to have a well-organized team with members cross-trained in every necessary skill.

I'd recently read a book called *Patriots* by an infamous survival blogger named James Wesley Rawles. A how-to book disguised as fiction, *Patriots* tells of a future in which inflation has made the dollar worthless, leading to social, economic, and government collapse. Fortunately, a group of eight friends has been training and stockpiling supplies for years—Just in Case. So they hole up in a compound in rural Idaho and, thanks to their military organization, survival skills, Christian values, and weapons expertise, successfully fend off looters, gangs, and even the United Nations.

The lengths they go to in order to accomplish this are not just extreme, they're inspirational. They build a 900-gallon diesel storage tank; a solar pump and 3,500-gallon water cistern; a

57-foot-high wooden tower for a wind generator; seven camou-flaged foxholes to ambush intruders; and bulletproof steel-plated doors and window shutters.

And that's just a small fraction of their preparations. They even add an extra fuel tank to their vehicles, which inspired me to look into doing the same.

I needed to start putting together a similar, if slightly smaller-scale, network. At the very least, I needed a few allies I could trust. A pack of wolves, after all, is a lot more powerful and dangerous than a lone wolf. Even Kevin Mason, the CERT fireman, had strategically made friends in his neighborhood.

Who did I have? No one.

"Now you do," Reeve replied when I shared my thoughts. "You can always come to us."

Although Reeve lived in New Jersey and would be difficult to get to WTSHTF, I appreciated his offer. It made me feel less alone. At least I now had someone with more experience than Spencer whom I could call for advice.

"But you can't come to us tomorrow," Alwood said, a cruel smile forming. Tomorrow was our final exam. "Because we'll be hunting you in the streets."

## LESSON 55

# A CROSS-DRESSER'S GUIDE TO COMBAT

It was nine A.M. on Sunday morning and I was in the backseat of a Range Rover, handcuffed again. This time, it was to another student. His name was Michael, and he was preparing to work in Iraq as a truck driver for Halliburton. He was trying to earn money, he said, to open a laundromat.

"Everyone has to wash their clothes," he explained, the dollar signs practically glinting in his eyes.

Reeve had driven us ten minutes outside downtown Oklahoma City, confiscated our bags, and left us handcuffed in the SUV in a parking lot in a desolate part of town. If we were caught anywhere in the city by Reeve and his cohorts—most of them bounty hunters and military trainers—we'd be put in restraints, thrown in the backseat of their car, and dropped off miles away to start all over again.

Luckily, I had internalized the first lesson of urban survival: planning. I'd spent the previous night locating supplies, hiding them in caches, and finding collaborators in the city. To make sure my bobby-pin pick wasn't confiscated, I'd made a thin slit in the seam of my shirt collar and stashed it inside.

I pulled it out and undid my handcuffs, then Michael's. Beneath the Range Rover floor mat was an envelope containing the first of several tasks we'd need to execute in downtown Oklahoma City to prove we'd learned to successfully navigate a dangerous urban environment. Our first assignment was to meet an agent wearing a black hat in the Bass Pro Shop in an area known as Bricktown and use persuasion engineering to get her to reveal our next mission.

Bricktown was a long walk away—especially since we'd get caught by bounty hunters if we took the main streets. Nearby, however, there was an Enterprise Rent-A-Car office; perhaps someone there would give us a ride.

The only customer inside was a young, muscular man in a large sleeveless basketball jersey. He was at least six inches taller than me and three times as thick. His face was crisscrossed with scars.

So I asked him for a ride.

"Our friends dropped us off here as a joke, and we have to make it back to Bricktown. Is there any way we can get a lift?"

"Do you got any guns or drugs on you?" he asked. That wasn't exactly the response I'd expected.

"No, definitely not," we reassured him.

"I'll give you a ride then," he grunted, "but I gotta warn you, if I'm pulled over by the police, I'm not gonna be nice to them."

I didn't know what he meant exactly, but it was scary as fuck. In that moment, I realized this wasn't a game. This was a real city, and this was real life.

Yet we followed him outside to a black Chevy Tahoe and climbed inside anyway. This, I realized as he drove us into town, was how people got killed. Evidently, in my mind, the law of conservation of energy had overruled the principles of common sense.

As he drove into town, he handed me his card. Underneath his name were the words CREDIT DOCTOR. "If you ever need your credit repaired, I can do it overnight—for the right price," he in-

formed me. He, too, was an urban survivalist of sorts, with his own method of beating the system.

He dropped us off in an alley in Bricktown where I'd cached a bag of disguises the night before. In a lecture on urban camouflage, Reeve and Alwood had taught us there was a certain category of people in cities called invisible men. If the city is a network of veins, invisible men are the white blood cells: they work to keep it clean. They're the janitors with bundles of keys on their belt loops, the alarm servicemen with clipboards and work orders, the UPS men hidden behind piles of boxes, and the construction workers with hard hats, safety vests, and tool belts. In these disguises, Reeve and Alwood said, we could walk unnoticed into almost any event.

However, since Alwood and Reeve had taught us these disguises, I knew they'd be looking for invisible men. But what they wouldn't be looking for was an invisible *woman*.

I slid under the back porch of a Hooters restaurant and found my bag of disguises. Miraculously, it was still intact in a small ditch in the rear of the crawl space where I'd cached it the night before. I grabbed the bag, climbed out, and entered a small corridor of shops above while Michael waited in the alley.

Inside, I found a public restroom and began my transformation. First I shaved my mustache and goatee in the bathroom mirror. Then I stepped into the stall and put on a flowery yellow cardigan I'd bought at Wal-Mart, after having seen a nondescript woman wearing a similar top.

I removed my cargo pants and replaced them with women's black dress slacks, then swapped my sneakers for yellow flats. Next, I pulled out a purse I'd stuffed with the rest of my disguise: hat, wig, sunglasses, clip-on earrings, and makeup Katie had recommended—face powder, mascara, and lipstick.

I left the stall to put on the hat and wig. Gazing at my reflection in the bathroom mirror, I realized, to my disappointment, that I didn't even make a good transvestite, let alone a passable woman. I hoped Katie's makeup tips would help.

I powdered my face, which helped conceal the faint outline of my freshly shaven beard. But as I was pulling out the mascara, the bathroom door swung open and a thick-necked college student with a crew cut and a striped button-down shirt stumbled in. His face was patchy and red, as if he'd been drinking.

He looked at me and slurred, "What the fuck are you doing?"

"I'm doing a class exercise," I blurted, hoping it would sound normal enough to calm him down. Then again, I was in a men's room in Oklahoma, dressed like a woman.

"What the fuck are you?"

I wasn't so sure I understood the question, but I tried to answer anyway. "I'm being chased by bounty hunters, and I need to dress like a woman so they don't recognize me."

He glared at me and knitted his brow. I tried to clarify: "It's for a course I'm taking on urban evasion."

In response, he opened the bathroom door and yelled into the corridor. "Hey, broheim, get a look at this."

Seconds later, broheim walked into the bathroom. He was bigger than his friend, and just as drunk.

"What do we have here?" he asked as soon as he saw me.

At this point, I was sure I was going to get my ass kicked.

With the two of them blocking the exit, I needed to put my survival skills to use immediately. Unfortunately, there were no locks to pick and no cars to hot-wire. And instead of my Springfield XD, I was carrying a purse.

I'd learned from Mad Dog that force respects greater force. So I ripped my hat and wig off in one motion, mustered as much toughness as I could, and told them coolly and firmly, "I'm in the

fucking marines. We're doing a drill in the city. Now back the fuck out before I get the rest of my battalion."

The thick-necked guy who started it all stared for a moment at my shaven head and then said, sheepishly, "I guess you are in the marines."

Thank God I hadn't attached the clip-on earrings yet.

I made a mental note to add another skill to my survival to-do list: hand-to-hand combat. I couldn't be a runner all my life. The only reason they were leaving was that they thought I was a fighter. I was reminded of something Tom Brown had said at Tracker School when teaching us to hunt: "A fleeing animal is a vulnerable animal."

After they backed out of the bathroom, I quickly changed into my jeans and tennis shoes again. Then I put on a military-green cap I'd bought, glasses, and a flannel shirt. With my facial hair gone, I hoped I'd be difficult enough to recognize. I'd learned my lesson: cross-dressing is not an urban survival tactic. It's an urban suicide tactic.

When I returned to the alley, my urban escape team was waiting for me. Michael had been joined by four locals I'd recruited by posting a bulletin on MySpace the previous night, asking for volunteers in Oklahoma City for a top-secret mission. (Evidently, there's not much to do in Oklahoma City on a Sunday afternoon.) Because the instructors had divided us into pairs, I hoped to escape their notice by moving in a larger group.

Sticking to alleys, parks, and industrial areas, we made our way to the Bass Pro Shop and safely carried out the first few missions. But then I made the mistake of leaving the group to grab another cache, which included a set of lock-picking tools. As Reeve had taught us, "Once you learn lock-picking, the world is your oyster."

I found the cache behind a pile of sandbags lying along the

banks of the city's canal. But as I made my way back to the group, I noticed a bounty hunter on a bridge above. He hadn't spotted me yet. But he would soon.

There didn't appear to be anywhere to hide or run—except for a door on the side of the bridge. I tried the knob. It was locked. I grabbed my lock-picking tools, found a pick with a flat underside, stuck it inside the lock, and raked it against the pins. There were five of them.

I selected a thin pick with an S-shaped end known as a snake and stuck it into the lock. At the bottom of the lock, I inserted a tension wrench. As I raked the snake along the pins, I pressed gently downward on the handle of the tension wrench. After a few minutes, the wrench began to turn. I pushed slightly harder on the wrench and, with a click, the door was open.

This class was better than my entire college education.

I needed to remember this wasn't a game. This was reality and it could have consequences.

After emerging fifteen minutes later, I rejoined my team and completed the remaining assignments, which mostly involved finding and photographing survival locations and items in the city: a water source, food source, daytime hiding location, safe place to sleep at night, easy-to-steal car, and an item that could be turned into a stabbing weapon.

This could just as easily have been a Fagin-like class for future career criminals. But like most governments, police forces, and armies, by calling ourselves the good guys, we had full permission to do any bad things we wanted—that is, until other people who thought they were the good guys felt otherwise.

While looking for water (available from several fountains) and food (available from edible plants and public ponds stocked with fish), I accidentally found several caches in the bushes made by homeless people. One contained a frying pan, the other a plastic bag with blankets inside. Between the cracks of the city, there was another world. And in that world, I learned, it was possible to live with no name and no money. I'd never thought of the homeless as survivalists before.

After completing our assignments, we reported back to Kevin. "How'd you get everything done so quickly without getting noticed?" he asked suspiciously.

Though I was worried he'd accuse me of cheating, I told him the truth: I'd recruited a scout and camouflage team on My-Space.

"I saw those guys, but I had no idea who they were. That goes down as one of the all-time great class stories."

I was relieved. Unlike wilderness survival, urban survival had no restrictions. Whatever worked was permissible. And that's why it appealed to me. After all, living like our primitive ancestors doesn't necessarily mean using sticks and stones. It means using every resource available and any means possible.

When I'd talked to Spencer after returning from Gunsite, we'd concluded there were only two scenarios to plan for: bugging in and bugging out. Thanks to Reeve and Alwood, I was finally ready to aggregate the skills I'd learned and conduct a trial run of the apocalypse to make sure I was fully prepared.

That is, after I called the Krav Maga center in Los Angeles and signed up for street fighting lessons. I wasn't going to get caught defenseless in a bathroom dressed as a woman again.

## LESSON 56

# EXAM #1:
# SHELTER IN PLACE

## DAY ONE

**12:39 A.M.**

Tomas calls and announces, "L.A. has shut down. Lights out, water off, gas off."

His timing couldn't be worse. I have a towel wrapped around my waist, and I'm about to step into the shower. But I must follow the rules. And the first rule is that Tomas will call me at a random time during the week to begin the three-day test, simulating a citywide emergency shutdown. So evidently I won't be showering for a while.

In fact, as far as I'm concerned, the outside world won't exist until Tomas comes by in three days and tells me the crisis is over. The goal of the test is not just to take care of myself during that time without utilities, but to make sure I have the necessary logistics and supplies to hole up for an entire month or longer, both here and in St. Kitts, in the event of a major disaster.

I grab the flashlight near my bed, walk to the fuse box, and shut off all the circuits. Then I find the master valve for the water line and turn that off. Next, I turn off my BlackBerry and unplug my land line. Finally, I grab one of my new survival tools, the Res-Q-Rench—a combination seat-belt cutter, window punch, gas shutoff, and pry tool—and turn off the gas.

In the dark silence that ensues, I feel calm and look forward to a deep sleep. Because I am prepared—or at least think I am—I don't fear the isolation. All I fear right now are the fish—which I've been practicing catching, gutting, and primitively cooking— spoiling in the freezer and stinking up the house.

I write on my troubleshooting list: "No fish."

**9:30 A.M.**

I bring my laptop into bed to write. I work for half an hour with no TV, no ringing phone, no Internet, no e-mail with a link to some YouTube video of a hamster licking its balls to distract me. I've never had a more productive thirty minutes. This little experiment might just be the best thing that's ever happened to me. At least that's what I think until I discover the first flaw in my three-day test.

I credit this discovery to my bladder after eight hours of sleep.

The problem is that I can't relieve myself in the toilets, because they may not have a last flush left. I can't go in my backyard, because it's small, I don't want to attract vermin, and, well, I just don't want to pee in it. And I can't use a bush outside, because there's a busy road in front of my house and I'll definitely be spotted urinating alongside it.

As I'm kneeling above my kitchen sink and unzipping my pants, worried that without water to wash it down the drain, my urine will reek up the house along with the dead fish, I remember the seventy-two-hour survival kit I bought from Nitro-Pak after 9/11. It remains untouched in my garage.

I zip up, find the duffel bag, brush off the dust, and open the top flap to discover a stack of water pouches and, below them, this gift from heaven:

It contains a cardboard base, a cardboard seat, and five biodegradable garbage bags. I assemble the Jungle Jon and carry it to a secluded spot in the backyard where my neighbors won't accidentally see me.

After finally emptying my bladder, I notice that the Jungle Jon has no lid. To keep the smell from attracting animals, I tie the ends of the bag together. I don't want to waste these bags. I'm going to need them.

As I walk back indoors, I'm reminded, for some reason, of when I was ten years old at overnight camp. Every Wednesday, the counselors took us to town so we could purchase a couple dollars' worth of candy. Most campers ate all their candy within the hour. But I hid the candy in my duffel bag to save for when I needed it most. Each week I stockpiled more sweets, until, at the end of the year, I had to throw most of my candy out. But I had

no regrets over the waste. Though I never got to eat it all, I always had candy when I wanted it.

Even at age ten, I was stockpiling supplies.

I write on my troubleshooting list: "More biodegradable toilet bags."

## 10:30 A.M.

I learned in CERT class that, in an emergency, I should first eat everything in my refrigerator that could go bad. Then I should move on to the freezer. And only then should I dip into the canned foods, MREs, and emergency supplies.

I check the refrigerator. There's milk, eggs, cheese, more butter than I can possibly consume in a day, tortellini, half a pizza, and leftover Chinese food.

I gather twigs and sticks from the side of the road, add a strand of jute twine to use as tinder, and build a fire outside so I can make an omelet. Then I check the refrigerator to figure out what to add to it. The carton of Chinese leftovers seems to be in immediate danger of putrefaction.

I look at the egg rolls inside. They're vegetables, meat, and eggs, basically. So I bring them to the fire, peel away the fried wrapper, and dump the contents into my omelet, avoiding the cabbage whenever I can.

The result: one of the most delicious omelets of my life. Necessity is truly the mother of invention.

I write on my troubleshooting list: "More egg rolls."

## 11:30 A.M.

My next challenge is one thing the survivalists never taught me: how to wash the dishes WTSHTF. Without a creek, lake, or pond nearby, I'm not about to waste my precious drinking water to scrub a frying pan.

As I stack the dishes in the kitchen sink, I realize there's water everywhere around me. Half the houses in the area have swimming pools, and most of my neighbors are probably at work. Though my house has a small pool of sorts, I drained it months ago because it was old and the plaster was cracking.

So I grab a waterproof plastic storage container from my garage, sneak into a neighbor's backyard, fill the box halfway with chlorinated water, and run home, sloshing liquid everywhere.

I'm only half a day into my survival test, and already I'm looting the neighbors. Maybe the gun nuts were right after all.

I write on my troubleshooting list: "Repair and fill pool."

**1:30 P.M.**

After eating, I return to the computer and enjoy an hour of uninterrupted writing time.

Until, suddenly, I feel a kick in my stomach. Now a gurgle. Then a push.

I try to ignore it. The last thing I want to do is sit on a piece of cardboard in my yard and take a dump in the blazing sun. I need to control myself. I can't go through this after every meal. I must hoard my shits like candy at camp and save them for when I want them.

Survival is easy until your body begins to make demands.

I cross out "More egg rolls" on my troubleshooting list.

**2:00 P.M.**

I walk outside to urinate but am horrified to see a small puddle leaking from underneath the Jungle Jon. I lift the bag to check for a hole, and there's water dripping from the plastic in multiple spots. Evidently, the bag is already biodegrading.

I take it outside and carefully throw it in a trash can. Then I grab a regular garbage bag and use it to line the Jungle Jon.

I never imagined so much of this journal would be devoted to

my waste functions. I'll never take a flushing toilet for granted again.

I cross out "More biodegradable toilet bags" on my trouble-shooting list.

**4:00 P.M.**

I eat the leftover pizza, then start the generator so I can recharge my computer. As I'm writing, Katie drops by with her sister.

Katie is wearing a tight pink T-shirt, a pair of low-rise jeans, and three-inch heels. She is the very apex of civilization.

"The air-conditioning isn't working."

"That's because I'm doing the three-day test now."

"It's too hot, baby. I'm going to Kendra's house."

"It's cooling off. I just opened all the doors." I don't want her to leave. I don't want to be alone. "Stay with me tonight. It'll be um, fun."

"Can I still use my curling iron?"

"I can put the generator near the bathroom, and we can use extension cords."

"What about going to the bathroom?"

"I have a little toilet I set up outside. Want to see it?"

"Does it flush?"

"No, it's just a little bag."

"I'm going to Kendra's."

I write on my troubleshooting list: "Get a more adventurous girlfriend."

# DAY TWO

**2:30 P.M.**

I've been so busy today I haven't been able to write until now. Living without conveniences takes time. I had to clean out the refrigerator and eat just about everything that could possibly rot. Then I gathered wood, built a fire, made soup, cooked fried chicken strips from the freezer, borrowed more washing water from the neighbors, cleaned the dishes and the grill, and added gas to the generator.

It would be a lot of work doing this for a four-person family. It seems that almost everything we call modern—everything we think separates us from previous generations—serves only a small number of purposes: to help us avoid pain, minimize work, and/or save time. Otherwise, we're not that much different from our primitive ancestors.

As soon as I finish lunch, my stomach cramps. There's no denying my body this time. Maybe the chicken was bad. I can barely untie the knot on the garbage bag fast enough.

The cardboard throne is surprisingly stable under my weight. And there's actually something pleasant about taking a dump while sitting in the sun. It's like being at a nudist beach. I can even get a tan at the same time.

If you ever see me with golden-brown skin but white marks in the shape of elbows on my knees, you'll know what I've been doing.

I write on my troubleshooting list: "Get Imodium. Just in Case."

**7:00 P.M.**

When night falls, I shut the doors to trap the heat in, bring all the tools inside, replace the batteries in my lights and flashlights, fill the lanterns with kerosene, and build a fire for dinner.

I take the last of my mostly defrosted fish out of the freezer, gut them, and stuff them with spices, lemon, and onion. Then I wrap them in tinfoil and place them on the coals of the fire. I'm

reminded instantly of the fate of the rebellious Iraqi villagers that the marine at Tracker School told me about.

While waiting for them to cook, I realize I need to shave. I bring a handheld mirror out to my dishwater container, set it up on the drying table, and start shaving.

When I finish, Katie comes by. Hopefully, she won't ask why my face smells like dish soap and chicken grease. She's with her friends Kendra and Brittany. She never arrives alone, because she's still too scared to drive.

"Romantic," Kendra says when she sees the house. Katie smiles. I can see this makes her more comfortable staying here. Maybe, instead of calling everything primitive, I can just call it romantic and get her more involved. Camping can be a romantic trip for two under the stars. Shooting guns can be a romantic fireworks display. Skinning an animal can be a romantic trip to the mall for a new coat.

"I know you don't like being here with the power out," I say, "but I'm making a romantic dinner right now and I'd love it if you could join me."

"What are you making?"

"I'm preparing some fish I caught."

"With your bare hands?"

"With a fishing pole."

"What about the bathroom?"

"If the bathroom toilet has one good flush left, it's all yours."

I lead Katie and her friends outside to show them the fire pit. The fish seem finished, so I remove them from the coals and un-wrap them. The skin peels off perfectly and the bone slips right out, leaving tender white meat that drops off in perfect flakes. Her friends stay to eat with us, and we feast on every bite in the glow of the lanterns.

"This is like the perfect Friday night," Kendra tells Katie. "You're so lucky. I have to stop dating these club losers and find a real man."

"Try checking the Santa Monica pier for fishermen," Katie replies.

Despite the joke, she's beaming ear to ear. She picks up a lantern and walks proudly to the bathroom.

I hope it flushes.

I cross "No fish" off my troubleshooting list.

## DAY THREE

**10:00 A.M.**

My refrigerator and freezer are nearly empty of solid food. No more eggs, no more meat except for two soggy chicken breasts I'm saving for dinner, not even an unmelted pad of butter. Though Katie wants cereal, the milk has gone bad. So I start a fire, heat up water, and make oatmeal instead.

"If we had a goat, we'd have milk for the cereal," Katie says, her head nestled on my chest as we curl up on the couch after breakfast. "It could be like our dog, but better than a dog, because it doesn't go around biting things. It just chews on shoelaces."

Since that fateful afternoon with Mad Dog, Katie hasn't stopped talking about Bettie. I suppose seeing her as a rug every time she enters the living room doesn't help much.

"I know," I tell her. "I looked into it after going to that permaculture place, but I don't know where to keep it."

"Maybe you could just keep it in the pool."

Farming is the first survival skill Katie has been supportive of. And I've actually found a person willing to sell me a beautiful Alpine goat. So maybe keeping it in the pool isn't such a bad idea. After all, a live dairy goat—perhaps paired with a male for

mating—will be better than a whole larder of canned food WTSHTF. And maybe it will help ease my conscience and fulfill the nurturing obligations of the circle of life that Mad Dog mentioned.

I cross "Repair and fill pool" off my troubleshooting list.

My neighbors are going to kill me.

## 1:00 P.M.

With Katie acting as a lookout, I take a covert plunge in the neighbor's pool in lieu of a shower, then dip into the rations for lunch—peanut butter, Nutella, and graham crackers. Afterward, Katie and I make love on the couch. It's the sixth time we've done it since last night. I could get used to this.

## 5:00 P.M.

Katie wants to make s'mores. I like that she's getting into the spirit of the test. We walk along the road together and gather twigs and dead leaves. I teach her how to build a fire, and we roast marshmallows on sticks together.

Since the fire's already going, I decide to make the last of my perishable food. I teach Katie—who's never cooked anything in her life but Pop-Tarts—how to prepare and season the chicken breasts. Despite her fears of fire, raw chicken, and work, she tosses the meat on the grill and cooks it.

"It's all wet and gushy and gross," she says.

"Am I going to get a disease from this?" she asks.

"I don't want to burn myself," she complains.

But she sees the project through. And when she's done, the chicken is soft and moist, with spices cooked deep into the meat. It pulls off the bone in scalloped shreds. It may be the best chicken we've ever had.

Though I can't wait to turn the power back on, I'm enjoying

this. I was never able to understand how people could live happily in the past without electricity and modern conveniences. Now I understand. They got along just fine.

I cross "Get a more adventurous girlfriend" off my troubleshooting list.

**7:00 P.M.**

I think of all the half-empty soda cans I've thrown in the trash, all the half-finished plates of food I've scraped into the disposal, and all the times I've said "no, thanks" when a waitress asked if I wanted a doggie bag. Now that my refrigerator and freezer are devoid of anything edible and every sip of water and morsel of food has become precious, I'm ashamed of my wasteful past.

I think of all the times I took long showers, left the lights on when I went out, ran water in the sink to cover up bathroom sounds, and maintained a house temperature of seventy-three degrees day and night. Thanks to this test, I've come to realize most of that consumption was completely unnecessary.

But because all those resources are there, because they seem limitless, because they're available at the press of a button or the flick of a switch, I—and most others—use them far too much.

Perhaps the root of the energy crisis is not our wasteful habits, but the ease with which seemingly unlimited power, gas, and water are available to us. In a normal three days, with Katie and her sister at the house most nights, I use 531 gallons of water. This week, I've used just a gallon. In a normal three days, I burn 121 kilowatt hours of electricity. This week, all my electricity has come from a gallon and a half of gasoline. In a normal three days, I use 2.28 therms of gas. This week, all my heat has come from dead leaves, sticks, and a ball of jute.

**10:00 P.M.**

I search through my rations for a late-night snack. I'm starving. Most of what I find is crackers, trail mix, and some canned food. Nothing seems substantial enough.

Then I notice my 9/11 safety bag in the living room and remember the box of MREs I'd also been talked into buying at the time. Supposedly they last seven years. And this just happens to be their seventh year.

I open an MRE. There's a packet of grape drink mix with absolutely no nutritional value, which I dump into a bottle of water to create some sort of weak Kool-Aid. The next package reads CRACKER, VEGETABLE ON IT. There's also a tube of grape jelly in the envelope, which I assume goes on the cracker. Together, they taste pretty good—and, due to some feat of army engineering, don't produce any crumbs.

I'm beginning to think I can live off these—until I open the next package, which is marked CHICKEN STEW. I squeeze it into a bowl, and it sits there brown and chunky like dog food. But I trust the army corps of food engineers, so I take a bite. I don't know if it's spoiled, if it's supposed to taste this bad, or if I need to heat it up or something.

As I'm trying to force it down, there's a knock on the door. It's Katie's sister, Grace, dropping by. As she sits down on the couch next to me, I notice she has a bag of takeout food in her hands.

I watch as she removes an entire pineapple from the bag. She pulls off the top of the fruit, and the smell of delicious Thai cooking fills the room.

"Do you want some?" she asks when she catches me staring at the steaming hot food she's spooning out of the hollowed-out pineapple.

"I can't. That pineapple shouldn't even be here. You're ruining my disaster simulation."

"Just try some. We won't tell anyone."

"I'm trying to see what it's like to live in a world without take-out restaurants."

"You're missing out."

"No, you are," I say, and take a bite of my gruel, trying to repress the gag reflex that follows.

I force the rest of the chicken stew down my throat, unwilling to let anything go to waste, then move on to dessert, which is a lemon pound cake that tastes surprisingly fresh after seven years in storage.

I write on my troubleshooting list: "Buy new MREs."

### 11:00 P.M.

Day three is coming to a close. Tonight or tomorrow, Tomas will come by and end the simulation. I feel like I've been on vacation, and I'm not looking forward to the phone calls, e-mails, and obligations waiting for me when I return to the plugged-in world.

Not only have I learned a lot from what was basically a simple test, but it's been one of the most romantic weekends Katie and I have ever had. We needed this.

If there's ever a citywide shutdown, I won't look at it as an inconvenience anymore, but as an opportunity.

# DAY FOUR

### 9:30 A.M.

"Baby." Katie is nudging me awake. "I want to shower."

Tomas should have been here by now. I check the door to see if he's left a note. Nothing.

I want to just turn the gas, water, and electricity on, but I must follow the rules. So I eat dry cereal, wash it down with warm water, turn on the generator, and write.

"When do I get to shower?"

"When Tomas comes."

"I'm going to Kendra's."

## 12:30 P.M.

I sit listlessly on the couch. I've gone through some beef jerky and tuna fish, and now I'm eating peanut butter and graham crackers again. I feel like life is passing me by. I could continue like this for weeks more, but I'm beginning to miss another thing the survivalists don't stockpile: variety.

There are no channels, and nothing's on. There are no restaurants, and no specials of the day. There are no websites, and no new episodes of *Homestar Runner*.

There's just me, in my house, with my stuff.

## 2:30 P.M.

Still no sign of Tomas. I start writing a comprehensive list of supplies I'll need to survive for a month, so I can keep stashes both here and in St. Kitts.

Of course, now that people know I'm stockpiling, the first thing all my friends will probably do when there's a disaster is run to my house.

I suppose they'll make an excellent source of protein.

Why do I keep making these jokes? Is there a half-truth somewhere in them?

I used to fear being the eaten. Now I fear being the eater.

**7:00 P.M.**

Still no sign of Tomas. I wish I could use my cell phone.

I'm going to walk to his house if he doesn't come soon. That's fair.

**10:00 P.M.**

This isn't fun anymore. Or romantic. It's lonely and irritating.

I have things to do, people to see, goats to order.

# DAY FIVE

**9:15 A.M.**

A loud, urgent knocking on the door wakes me.

I walk downstairs to find Tomas, looking repentant.

"Did you turn everything back on already?" he asks.

"No. I was following the rules."

"I'm sorry."

"Well, is the emergency over?"

"Yes. Sorry. It completely slipped my mind."

I hate him right now. But I'm also grateful for the lesson he accidentally taught me. It may be fun to have a vacation from modern conveniences, but it's hell to live there.

I write on my troubleshooting list: "Fire Tomas."

# HOW TO ESCAPE FROM FLEXICUFFS

YOU'VE ALREADY READ ABOUT ESCAPING FROM HANDCUFFS USING A BOBBY PIN. BUT THERE ARE OTHER WAYS THE ENEMY MAY TRY TO RESTRAIN YOU.

IF YOUR HANDS ARE BOUND WITH ROPE, YOU CAN ESCAPE USING THE STRUGGLE-TWIST-WRIGGLE METHOD. SIMPLY MOVE YOUR HANDS UP AND DOWN QUICKLY WHILE ROTATING YOUR WRISTS IN HALF-CIRCLES. EVENTUALLY, THE ROPE WILL LOOSEN ENOUGH SO YOU CAN SLIP OUT.

IF YOU'RE BOUND WITH DUCT TAPE OR THIN FLEXI-CUFFS, MAKE SURE YOUR HANDS ARE IN FRONT OF YOUR BODY. PULL THE RESTRAINTS AS TIGHT AS POSSIBLE AGAINST YOUR WRISTS, THEN RAISE YOUR HANDS UP. NOW BRING YOUR ELBOWS DOWN AND BACKWARD AS FAST AND HARD AS YOU CAN, SO THAT YOUR WRISTS STRIKE YOUR RIBS AND THE PRESSURE RIPS THE RESTRAINTS OFF.

ANOTHER WAY TO ESCAPE FROM FLEXI-CUFFS IS TO WEDGE A NEEDLE, NAIL, OR BOBBY PIN BETWEEN THE TEETH OF THE CUFF STRAP AND THE PLASTIC ROLLER LOCK KEEPING IT IN PLACE, THEN JUST PULL THE CUFF OPEN.

BUT THE BEST WAY IS TO KEEP YOUR SHOES LACED WITH PARACORD. THEN REMOVE ONE OF YOUR LACES, AND HANG IT FROM THE PIECE OF PLASTIC CONNECTING THE TWO SIDES OF THE CUFF.

TIE A LOOP ON EACH END OF THE SHOELACE, AND STICK ONE FOOT IN EACH HOLE. THEN MOVE YOUR FEET UP AND DOWN, AS IF YOU'RE RIDING A BICYCLE, IN ORDER TO CREATE FRICTION BETWEEN THE PARACORD AND THE PLASTIC.

USE LONG STROKES, AND EVENTUALLY YOU'LL BURN THROUGH THE PLASTIC. IF THE PARACORD BREAKS FIRST, USE YOUR OTHER SHOELACE OR ROTATE YOUR WRISTS TO TEAR THROUGH THE REMAINING PLASTIC.

*TO BE CONTINUED...*

## LESSON 57

# EXAM #2: BUG OUT

## LAND

WTSHTF, some people won't have to worry about finding a bug-out location. If they don't live in an isolated area already, they at least have a close relative living on a farm, raising a family in an outer suburb, or renting a summer retreat.

But I come from a long line of committed urban dwellers. They live in the high-rises of Chicago, the working class boroughs of Boston, the hillside homes of San Francisco—basically everywhere you don't want to be when the system collapses. Consequently, since restocking my supplies after my three-day test, I was well-prepared to bug in; but I still lacked a failsafe plan to bug-out.

"If you're going to live in the city, the key is getting out ahead of time," said Kevin Reeve from urban evasion class when I called him for advice. "You need to keep your finger on the pulse and look for tripwires. If Israel attacked Iran, for example, I'd be on the next plane to St. Kitts. If you get stuck in L.A., your next best bet is to shelter in place. Bugging out would be your last resort."

On his next trip to California, Reeve took a few days off work to help devise my escape. Evidently, he meant what he said over dinner in Oklahoma City when he told me he'd be my network.

Reeve and I spent the morning studying topographical maps

of California, assessing the survival potential of dozens of lakes and reservoirs. Our goal was to choose three different locations and escape routes, so that no matter what type of disaster struck, I'd have at least one getaway option.

That afternoon we tested our first bug-out plan, and rode a motorcycle through the hills and fire roads to Malibu Lake thirty miles away. En route, we checked for possible obstacles, traffic choke points, and roads wide enough so that I could weave to avoid shooters.

"This area actually has everything you need to survive," Reeve informed me when we arrived at the lake. "There's water, fish, edible plants, animal tracks, and yucca for cordage. You could hide in the cottonwoods here for weeks."

As we scouted the area, he plucked a yucca leaf and showed me how the pointy tip could be used as a needle to suture wounds. Then he picked up a stick and poked at a lump of nearby poo to see what type of animal had made it.

"It's composed of cereal," he informed me. "Must be a dog."

I added poo reading to my survival to-do list.

Then I scratched it out. Some skills were better left to the experts.

## AIR

In search of a second, more remote bug-out option, I called Mad Dog for advice.

"You need to get yourself an autogyro," he said firmly, as if it were as essential as a good knife. "Then fly over all the shit rather than navigate through it. If you get an autogyro with pontoons, you can land on a mountain lake that has no road access. You could live there forever with just a seven-pound tent, a little fuel, an axe, a saw, and a fishing pole."

It sounded like the perfect way to isolate myself from the chaos and danger of a collapsing society. So I started looking into gyrocopters like this one:

The problem was that gyrocopter kits cost between $10,000 and $30,000, and my survival expenses had already put me in debt. In addition, I had nowhere to store it, so I'd have to rent extra space somewhere. I inspected helicopters as well, and though they also seemed ideal for bugging out, a decent used one cost well over $150,000.

So I chose a more affordable option: United Airlines. Obviously, a commercial jet wouldn't be a feasible escape option WTSHTF, but for now, it was the quickest way to find a second bug-out location far from the smog of Los Angeles.

To make sure I learned something from the experience, I gave myself a new test that probably would have killed me six months earlier: I decided to make sure I'd completely shed my addiction to the system and could survive three days in the wilderness with nothing but a knife and my wits to sustain me.

In case I accidentally ate California parsley surprise and died, I brought Reeve along to keep an eye on me. As additional backup, after I arrived I buried a cache of survival supplies.

Though I'd spent months preparing, the test was nothing like what I envisioned. Nature has a way of defying even the best-laid plans.

I didn't have enough daylight to build a debris hut when I arrived, so I had to sleep against a tree and cover myself with a mound of leaves for warmth.

It took half a day and both of my shoelaces to make a bowdrill and start a fire.

It took three days and a three-mile hike before I found a single edible plant.

I didn't even come close to hitting an animal with my throwing stick. And I couldn't set traps because they're illegal in national parks.

It took hours to purify water, because open fires were prohibited and I had to build a small, hidden scout fire.

And when I finally caught an animal with my bare hands—a salamander with translucent skin, bulging eyes, and a long, slimy tail—I didn't have the stomach to eat it.

But despite my discomfort and salamander squeamishness, I survived. I lived off the land for the first time in my life. It was one of the most liberating feelings I'd ever had. By the time I left, I knew I'd be able to stay in the wilderness as long as I needed. I wouldn't be comfortable, but I'd be alive. I couldn't believe how far I'd come since my snivelly nights at Tracker School.

Though I didn't end up using the cache, I left it where I buried it—not just in case I decide to bug out there in the future, but in case you do.

If you ever need it, this is what it looks like:

I'd tell you exactly where it is, but then everyone else reading this could also get to it. So I've compiled eight short clues, each hidden in one of the illustrations in this book. If you can decipher them, they'll lead you directly to my emergency cache.

When you retrieve it, email me at StSlimJim@gmail.com to receive an additional reward that won't fit in a hole in the ground. You may also contact me there to find out if it's already been unearthed.

If you can, bury something else there for other treasure-hunters. In the meantime, until I can save up for a gyrocopter kit, assemble it, and learn to fly there, I have other bug-out plans.

## SEA

Wind and salt spray whipped against my sun-beaten face as the sailboat picked up speed, dipping its rails into the deep blue of the Pacific and bringing me closer to the final location in my bug-out simulation. I couldn't imagine a more ideal and exhilarating way to escape the city WTSHTF. There'd be no traffic jams, no roadblocks, no need for fuel, and no looters out here, unless a small pirate subculture suddenly developed.

And fortunately, Reeve was in the process of telling me how

to deal with that very contingency. "If you're attacked by pirates, you can sink their boat with your Remington shotgun. Just remove a slug from the shell casing and drill a quarter-inch hole in the slug. Then insert a .22 short round backward and replace the slug. When fired at the water line of most boats, the round will blow a hole that can sink it."

We were on our way to Catalina Island, twenty miles due south of Los Angeles. With a population of four thousand, the island would be relatively safe from rioting and violence.

An inflatable dinghy, stocked with supplies and stored near the ocean, would at least be more affordable than anything in the copter family. With a fishing pole, I'd have easy access to food. With a desalinator, I'd have a lifetime supply of drinking water. And with a sail kit, I'd have an inexhaustible source of energy.

"You know this is pointless, right?" Katie's voice interrupted my scheming. I'd brought her on my final test run because if I ever bugged out, I'd be taking her with and I wanted her to be prepared.

"What do you mean?"

"You're just going to Catalina to dig holes in the ground for no reason."

"Do you mean the cache?" She shrugged. I'd brought along an Army ammunition can loaded with freeze-dried food, water, and other survival essentials. "The point is to bury supplies. This way, as long as I can sail to Catalina, I can always survive there."

She dropped her head into my lap, stretched her legs along the sailboat bench, and fell fast asleep. She was tired from the Dramamine she'd taken for seasickness, and in a generally cranky mood. Not only was she worried the boat would hit a rock and sink, but she hadn't eaten dinner the night before or breakfast that morning.

She reminded me of the boy in Cormac McCarthy's *The Road*, whose innocence and compassion constantly interfere with his father's attempts to keep him alive. I wondered whether the compassionate would be the first to go WTSHTF or if they would outlast the remorseless, which I was fast becoming.

Reeve reached into a paper bag and pulled out three tuna sandwiches we'd packed for lunch. I woke Katie and handed her a sandwich, but she blinked groggily at it, then fell back asleep.

By the time she woke up, the sandwich had spoiled in the sun. On the horizon, the outline of Catalina Island began to take shape. She'd be able to get food there soon enough.

Three hours later, we dropped anchor at the port of Two Harbors in Catalina, ten miles from the main city of Avalon. Reeve and I climbed into the dinghy, while Katie stayed on the boat to nap until dinner.

Racing against waning daylight, Reeve and I motored along the coast until we came to a deserted stretch of beach. After finding a well-concealed site in the hills above the shoreline, we took compass bearings and then, like pirates burying treasure, dug a deep hole to drop the cache into.

An hour and a half later, we returned to the sailboat wet, exhausted, and elated. Bugging out was a lot more fun than bugging in. I looked forward to ending a hard day of work with a hot meal at Two Harbors' one and only restaurant, famous for its seared, slow-roasted prime rib.

While Reeve napped below deck, Katie and I took the dinghy ashore. By now, her hunger had turned from craving to illness.

"My stomach feels like it's being tied in knots," she complained. "I would give anything for buffalo chicken strips and hot sauce and onion rings right now."

Thankfully, the restaurant was open. Though no one was sitting at the tables, there were a few patrons at the bar. "Can we see a dinner menu?" I asked the bartender.

"The kitchen is closed," he replied brusquely. "There weren't any customers, so the cook went home early."

Katie's face fell. She asked if there was another restaurant, a pizza place that delivered, or a convenience store. As the bartender responded to each question with a curt no, her eyes began to blaze.

"What if we go back to the boat?" she asked me, panic setting in. "Is there any food there?"

"Um, we buried it."

"Can we go get it?"

"Sure, I guess, but it will take a while to go back and dig it up." I was so focused on preparing for future catastrophes that I'd forgotten to plan for the present. But at least Katie now understood the importance of digging holes in the ground.

"I can't wait that long," Katie seethed. She scanned the bar for someone with chips or a sandwich. "I'm in such a hateful mood. I would viciously kill someone for food right now."

Maybe Katie wasn't as much like the boy in *The Road* as I'd thought. The hunger in that book was fictional. Her hunger was real. "This is exactly why I've been learning to shoot guns," I told her, hoping she'd finally realized the importance of being self-reliant.

"Because one day you'll be hungry and need to murder someone for food?"

"No, so I can defend myself from people like you."

"Well, if I had a gun, I would literally go on all those boats out there and tell them, 'Give me your food now.'" She glared at the bartender irritably. "I don't like being hungry."

There were moments when I doubted whether all the time I'd

dedicated to learning survival at the expense of movies, parties, writing, friends, and vacations was worth it. But not anymore. If even Katie could be turned violent when hungry, what would happen to a whole nation of hungry people?

She looked up at the bartender, forced a smile, and pleaded, "Do you have any bread?"

"I guess," he answered, "but it's cold."

"You don't understand how hungry we are right now."

It had only been one day without food, and already we were begging for a scrap of bread.

The bartender walked away and returned with two rolls. After wolfing one down, Katie turned to me. "I'm sure he could get us all kinds of food from the kitchen." She paused to shoot the bartender another dirty look. "If all of us were stuck here, he's the first person I'd choose to eat because he's so mean."

In that moment, I realized just how easily the snap I feared could be triggered. It wouldn't necessarily take three days without food, as Reeve had taught. It could actually take less than a day. Because it wasn't just the lack of food that was making Katie break down—it was the lack of hope. Without a peaceful solution to her hunger in sight, she'd been forced to devise an alternate plan. And when you combine hopelessness, desperation, and pain—and then dangle a remedy just outside the reach of social acceptability—the sum is often violence.

Fortunately for the bartender, when we returned to our boat, we found a few crackers to sustain us. After a fitful sleep, we sailed early the next morning to Avalon and refortified ourselves with a hot breakfast.

Perhaps bugging out wasn't as fun as I'd originally thought. Without a well-stocked retreat, hiding on Malibu Lake or Catalina Island wouldn't be easy. If I couldn't eventually get to a better-fortified location, I'd waste away to nothing.

"Maybe I should start saving up to build a retreat somewhere nearby," I told Reeve, "just in case I can't get to St. Kitts ahead of time."

"That's probably a good idea. I'd look into Northern Idaho around Coeur d'Alene or Washington around Spokane. There's cheap land, a low population density, lax gun laws, and a minimum of strategic targets."

As soon as I heard the word Spokane, a memory I'd forgotten about came roaring back to me. My father once told me that in the early 1900s, my great-grandfather had sent his children one by one from Germany to a rural area in Washington State, near Spokane.

When I was a child, I'd even visited a cabin there with my grandparents. I remembered tagging along as one of my cousins picked huckleberries, caught fish on a quiet lake just outside the house, and took photos of a moose that was wading in the water.

It was a survivalist's paradise.

I grabbed my phone and dialed my parents.

"What ever happened to that cabin in Washington grandpa used to have?"

"We're still part-owners," my father said.

"Is it okay if I use it as a retreat in case anything bad happens?"

"I don't see why not."

"Why didn't you tell me that before, when I was looking for a good place to escape to?"

"We never thought about that," my mother cut in. "You're supposed to be the smart one."

I pressed them for more details about the cabin. It had fishing tackle, rowboats, firewood, and a root cellar for food storage. It seemed ideal.

"I'd be glad to come up there with you at some point and help

you stock it," Reeve offered. "We'll cache gas along the route to make sure you can get there."

In addition to my three bug-out locations, I now had a real retreat.

However, as I flew home, I didn't feel as confident as I thought I would. Since the blackout in St. Kitts, I'd absorbed what felt like a lifetime of information. But even though I'd just successfully tested most of it, in my heart, I knew that if the S really HTF, these simulations weren't enough to guarantee my survival. I was still lacking something important, besides my discouragingly delayed second passport. Something that would make all the difference between life and death. And it would be the most difficult thing to find: experience.

## PART FIVE

# RESCUE

If you wish to become like the gods and live for days without end,
you must first possess the strength of a god.
Even though you are mighty I will show you that,
like all human beings, you are weak.

—*Gilgamesh*, Tablet XI, 2100 B.C.

## LESSON 58

# THE PRESIDENT'S PLAN TO SAVE YOU

When the next major natural disaster or terrorist attack occurs, here's what will happen:

The fire department, police, and ambulance companies will race to the scene. The most qualified officer will become the incident commander. He will establish an incident command post, or headquarters. Until he is relieved, he will be in charge of the entire operation, unless there are multiple incidents (such as bombs detonated in different locations), in which case he and the incident commander at each location will report to a single area commander.

In the meantime, the injured will lie there bleeding, screaming for help. But that's okay. Without an organized plan, chaos would ensue and more lives would be lost.

The incident commander will then appoint a public information officer to deal with the media and the community; a liaison officer in charge of getting assistance from other agencies and jurisdictions; and a safety officer to make sure emergency workers aren't putting themselves or others in danger.

If the situation is so catastrophic that multiple agencies are involved or it crosses jurisdictional lines, the incident commander will set up a unified command, with incident commanders from each major agency running the operation from a single shared command post.

People are still bleeding. But that's okay. They'll be taken care of shortly. Except for the ones who are in shock and massively hemorrhaging. They'll probably die.

The incident commander will then set up four departments, with officers in charge of each: planning, logistics, operations, and finance/administration.

The operations section chief will organize a number of different teams—usually a rescue team to extricate victims, a medical team to triage and treat them, a fire team to extinguish any blazes, and a hazmat team to handle hazardous materials and those exposed to them. He will also appoint a staging area manager, who will send arriving rescue workers to join the appropriate team.

Finally, rescuers will begin working to make the area safe and extricate victims. If they can reach any survivors without endangering themselves, they'll begin triaging them, treating them, and transporting the seriously injured to nearby hospitals in helicopters and ambulances.

If the disaster is too big for local agencies and resources to handle on their own, the incident commander will alert the regional emergency operations center, which will activate the state operations center, which will brief the governor, who will request a joint preliminary disaster assessment from the Department of Homeland Security and, if appropriate, request a presidential declaration of a state of emergency through the FEMA Regional Administrator, who will evaluate the request and forward it to the national FEMA Administrator, who will report to the Secretary of Homeland Security, who will pass the recommendation on to the president.

If this seems confusing, that's because it is. By now, anyone in critical condition who wasn't lucky enough to receive care via

local resources is dead. As for everyone else, they're stranded, starving, and sick, wondering when more help is going to arrive.

The president will then declare a state of emergency and activate the National Response Coordination Center, which will make an overall plan, notify the involved federal agencies, and deploy emergency response teams. All these agencies will then create and staff a joint field office to coordinate government resources and, finally, deliver federal assistance.

It's all been nicely worked out by the government in a system every emergency responder is trained and tested in: the National Incident Management System. As a testament to the planning power of the human mind, the many agencies and inner workings of the country's emergency response apparatus are awe–inspiring in their sophistication and complexity.

But just because there's an organizational system doesn't mean that people will be organized.

They're still human beings.

When disaster strikes, even after the delays caused by the implementation of the federal plan, arriving teams often end up asking each other if they know what's going on and what they're supposed to do. As victims lay trapped in the wreckage, emergency responders are arguing over whether they can take responsibility for freeing them. While people are dying, officers are trying to determine what department has jurisdiction over the ground they're dying on. As requests for assistance are traveling up the federal ladder, the death toll is rising as agencies argue over their responsibilities, try to locate key employees and politicians, and assemble further committees and teams.

In a large-scale disaster, it may take three to fourteen days before the system begins working as it's designed to. Thus, the National Incident Management System is not necessarily a means of turning disaster response into science, but a new avenue of debate and miscommunication.

I know because I'm now part of this system:

### Emergency Management Institute

**FEMA**

This Certificate of Achievement is to acknowledge that

**NEIL STRAUSS**

has reaffirmed a dedication to serve in times of crisis through continued professional development and completion of the independent study course:

**IS-00200.a**
**ICS for Single Resources and**
**Initial Action Incidents**

*Issued this 27th Day of December, 2008*

Cortez Lawrence, PhD
Superintendent

Los Angeles County Disaster Management Area G

presents this

## Certificate of Training

to

### Neil Strauss

upon completion of NIMS ICS-100 and IS-700 Training Program

October 4, 2008

*Mike Martinet*, MS, CEM
Area G

A lot has happened in the last few months. Let me fill you in.

# ICELAND IS THE NEW CARIBBEAN

Maybe it was when the price of a gallon of gas broke the $5.00 mark in some parts of the country.

Maybe it was when Bear Stearns became the first brokerage house to be rescued by the government since the Great Depression.

Maybe it was when IndyMac became the fifth American bank to fail in recent months.

Maybe it was when the government gave customs agents authority to confiscate, copy, and analyze any laptop or data storage device brought across the border.

Maybe it was the housing market crashing, the unemployment rate rising, the stock market slumping, the credit card debt ballooning, the cost of living increasing, the national debt skyrocketing, the public-high-school drop-out rate hitting fifty percent in major cities, and Afghanistan and Iraq showing signs of becoming the longest wars in U.S. history.

Maybe it was the unshakable sense that the worst was still to come.

But I was no longer alone.

It was a hot summer, and pessimism hung thick in the air. Most people I talked to felt as if they were inching closer to some darkness they couldn't understand, because they'd never experienced it before and didn't know what it held.

The grimmer the news became, the more I harassed Maxwell in St. Kitts. I began to worry that I'd have to start looking for a second passport from scratch, which wouldn't be as easy now that the idea of an exodus was starting to spread.

Even Spencer's housemate Howard, who had once made fun of us for taking precautionary measures, was now looking into Caribbean islands. As it turned out, he would beat all of us there when his company collapsed and he had to hide from possible indictment.

"I'm so glad we started preparing ahead," Spencer told me over dinner at the Chateau Marmont, where he was staying in Los Angeles.

Having struck out with the Swiss, I took Spencer's advice and opened an account with a Canadian bank that had a branch in St. Kitts. Since both Canada and St. Kitts are part of the British Commonwealth, he'd explained, I would have easy access to my money if anything happened in America. Unfortunately, in the process, I discovered that keeping international accounts secret is now illegal: the IRS requires Americans with over $10,000 in foreign accounts to file an annual report disclosing not just the amount of money and the banks it's kept in, but the account numbers.

Meanwhile, Spencer was moving forward with his ten-year plan. He started an Internet business in Singapore, enabling him to open a private banking account in the country, which he claimed was fast becoming the new Switzerland. Though he hadn't gotten his St. Kitts passport yet either, Spencer had done more research into buying an island.

"I'm looking at islands in the north, around Iceland, because no one will think of looking for anyone there," Spencer said, his thick lips spreading into a self-satisfied smile. "If I can get some

other B people to go there with me, we can build underground homes and use geothermal energy."

"What about your submarine?"

"It's a great way to move between islands undetected, but we're running out of time. We need to move faster. This is only the beginning."

"How bad do you think it's going to get?" Spencer seemed to understand the economy at a higher level than most people did, perhaps because he knew so many of the people who ran it.

"I don't think the whole country's going to collapse, but we're looking at the worst economic disaster in America since the Great Depression. What I'm also concerned about is the increase in violent crime that's going to accompany this."

Everywhere I went that summer, the demon of Just in Case seemed to follow me, growling in my ear louder than it ever had, its jaws terrifyingly close to my jugular. I'd learned so much, changed so much, tested myself so much. It now was time to stop preparing, turn around, and face the demon—and my fears—head on.

And a musician would lead me there.

## LESSON 60

# WHAT ABOUT LOVE?

**W**hat I worry about," Leonard Cohen said as we sat in the spartan living room of his Los Angeles apartment, "is that the spirit will die before the country dies."

His words were pithy and poignant, like those of his songs.

Though some turn to politicians, priests, or parents when seeking answers to questions about the world, I'd always turned to music. And, more than anyone, Leonard Cohen was politician, priest, and parent to me. The beauty of his words was one of the things that inspired me to write.

A mutual friend, a record executive named David, had taken me to see him after hearing me talk about survival. He thought I'd find a kindred spirit in Cohen. As the economy imploded and the 2008 elections neared, it seemed like most Americans were becoming kindred spirits in anxiety. Except David.

Soft-spoken and reclusive, David spent more time reading spiritual books than newspapers. Unlike Cohen and I, who fell into a class of people known as negativity avoiders, he was a positivity embracer.

"But what about the protests against the Iraq war?" David challenged as we sat in Cohen's living room. "That was the first time anyone protested a war before it happened. Maybe that's a good sign of a consciousness awakening."

"There are forces of evil in this world that are too great," Cohen responded. His wrinkles were deep-set under gray hair, framing hazel eyes that shone with the vigor of a twenty-year-old student and the intelligence of a hundred-year-old monk. "The war was an inevitability."

He took an unhurried sip of tea. As for the protests, he continued, "You become what you resist."

Cohen stood up to put a pot of bean soup on the stove. He was seventy-three, and his manager had recently been convicted of stealing most of his retirement fund, leaving him nearly broke and mired in lawsuits.

Just miles to the south of him, Cohen said when he returned to the living room, were some of the most dangerous gangs in L.A. And when their anger reached critical mass, he worried that they would spill out of their neighborhoods, and the police would be too busy trying to protect their own families to stop them from terrorizing the city.

"My fear," he concluded, "is the social contract breaking down."

"The snap?" I asked.

"Yes, the snap." And right then, I knew he was one of us: a Fliesian.

I told Cohen about the preparations I'd made and the survival skills I'd learned. "Look at me," he said after I finished. "I'm not made to survive. I could learn to shoot guns, but I don't have any experience in the street. I'm not a fighter."

"What about love?" David interrupted. "You don't have to hide from violence or fight against violence. You can embody love." I envy positivity embracers for the ease with which they glide through life. They die happy dreaming of heaven. We die miserable worrying about hell.

Cohen's answer came quickly, as if he'd already come up with it decades before, perhaps when he was playing Vietnam War protest concerts. "That's what the enemy wants, because then it's easier for them to conquer you. The people who survive persecution do so because they are strong."

He then quoted a Zen aphorism: "The lotus that blooms in the water withers when it comes near to fire. Yet the lotus that blooms from the midst of flames becomes all the more beautiful and fragrant the nearer the fire rages."

I had bloomed in the water, as had most Americans raised in the late 1970s, '80s, and '90s. Consequently, as Cohen put it, we were weak. That's why the events of the last few years had sent me into such a panic. Thanks to Tracker School, Gunsite, and Krav Maga, I was getting stronger. But if I truly wanted the ability to survive, I needed to expose myself to the heat, to harden myself, so I'd be inoculated when the flames started licking my heels.

Perhaps the main reason I was even on this journey was that I didn't want to take a relatively sheltered life for granted. As a teenager, I'd been subjected to so many books, classes, movies, and comic books about sex and violence that it seemed strange neither of those two things had happened to me yet. Eventually, after a lot of work and anticipation, sex came into my life. But what about violence? Would I be ready for that eventuality? Because, unlike sex, I might only get one chance to face it.

When I tuned back into the conversation, Cohen was discussing a book he'd read about Auschwitz. Some scholars, he was saying, wonder why the Jews didn't rush forward and try to overpower the handful of machine gunners about to shoot them when they were being led to mass graves.

"Because they knew that, even if they succeeded, they couldn't escape?" I replied, attempting to contribute.

"It's because that's not what they wanted to do," Cohen replied. "They wanted to reflect on their life and prepare to die." He paused and slowly picked up his teacup. "And that's what I'm doing. Preparing to die."

"How does one prepare for that?" I asked.

Cohen never answered. He stood up to check on his bean soup, then led us to his office to play a song he'd recently written. It was a long list of horrible things that are going to happen, followed by the plea, "Tell me you still love me."

It was unclear whether he was pleading with a woman or with God.

I drove home from Cohen's house feeling more anxiety than I had in a long time. I couldn't get his lotus metaphor out of my head. It pointed to a gaping hole in my survivalist training. Except for being mugged twice, I'd never been through any real disaster, trauma, or emergency. And outside of my Oklahoma City bathroom experience, I'd never had to run or fight or talk or struggle to save my life. I'd never even seen a dead body before. So if I was under stress and my life was in danger, I had no idea whether I'd be able to effectively use the skills I'd learned—or if I'd just panic, freeze, and mess my pants.

I spent the next week thinking about ways to expose myself to stress and danger, considering everything from paintball (not real enough and ultimately harmless) to the military (too real and possibly fatal) to injecting myself with an EpiPen to see if I could handle the adrenaline rush.

"Maybe you should try jumping in front of a train and then jumping away right before it hits you," Katie suggested when I asked for ideas. "Or what if you walked through Compton without your gun, and you were wearing a shirt that said 'Bloods' on it? That would be high-stress."

"Do you really want me to do those things?" I asked.

"No, I don't. Because then you're going to die."

Low on ideas, I decided to turn to the experts I'd met for advice.

"The best cure for stress is repeated exposure," Kelly Alwood from urban evasion class suggested when I called. Unlike me, Alwood had a father who was a Green Beret and had been teaching him to toughen up since he was six years old. My father had mostly taught me about jazz.

"So what do you do if you have no exposure, outside of the regular stresses of life?"

"I'm taking an EMT class now. It's the best emergency medical training you can get. And it's great for stress inoculation, because as part of the class, you have to ride in ambulances and respond to 911 calls."

My heart raced as he spoke. There were people in Los Angeles who put themselves in the face of danger every day: firemen, police officers, paramedics, search-and-rescue teams. I needed to seek them out and join their ranks.

Not only would I get the experience I was looking for, not only would I get a uniform and badge that would get me past roadblocks when escaping the city, not only would I get keys to the back fire roads, not only would I be exposed to life-and-death situations, but I'd have the best, strongest network available: the system itself.

Maybe Leonard Cohen was right. You become what you resist.

## LESSON 61

# THE ODDS OF DYING HORRIBLY

In the event of amputation of the penile shaft, whether partial or complete, an EMT should "use local pressure with a sterile dressing on the stump."

If during "particularly active sexual intercourse, the shaft of the penis is fractured or severely angled," an EMT should immediately transport the patient to a hospital for possible surgical repair.

If the foreskin of the penile shaft is caught in one or two teeth of a zipper, the EMT may attempt to unzip the pants. But "if a longer segment is involved or the patient is agitated, use heavy scissors to cut the zipper out of the pants . . . Be sure to explain how you are going to use scissors before you begin cutting."

These valuable life-saving tips come from the book *Emergency Care and Transportation of the Sick and Injured*—1,296 densely packed pages detailing nearly every medical and traumatic emergency that could possibly happen to a human being in a civilized country, from strokes to gunshot wounds to zipper mishaps. It was one of the best, most concise, most information-dense survival resources I'd read since *FM 3-05.70*, the U.S. Army's survival field manual.

It was also my textbook at the California Institute of Emergency Medical Training, where I was studying to become an EMT.

Alwood had steered me in the right direction. The course was exactly what I needed. In fact, the very first class was about handling stress.

"Are you going to see people die?" Matt Goodman, the small, hyperactive head of the institute yelled at us.

The class listened in silence.

"Yes, it's actually part of the job. Is this stressful?" He paused and waited for a response. A few students murmured in agreement. I was so out of my element that I had no idea whether my classmates were silent out of shock or boredom. "It's definitely stressful. The truth is, we see people die. But you don't have the luxury of screwing up just because you're under stress. You cannot let it affect you at that time. Family members may lash out at you. Don't try to win them over or change their minds. Stay calm and let it roll off your back."

I felt like an imposter. The student on my right wanted to work with the fire department. The girl on my left was applying to medical school. The guy in front of me was part of a local search-and-rescue-team. And me? I was practicing for the apocalypse.

Somehow I'd crossed a line in my pursuit of safety. I was no longer taking courses teaching me things to do Just in Case. I was actually going to be doing these things.

In subsequent classes, Goodman drilled us on how to recognize and, when possible, treat everything that can go wrong with the human body: heart attacks, car crash injuries, strokes, allergic reactions, seizures, poisonings, gunshot wounds, broken bones, burns, drowning, hypothermia, heatstroke, drug overdoses, childbirth complications, lung diseases, diabetic emer-

gencies, snake venom absorption, lightning injuries, and scores of other diseases, accidents, and mental disorders.

And, though we were taught how to recognize and deal with symptoms related to nearly every weapon of mass destruction conceivable, we also learned that there were bigger things to be scared about than a jihad-bent teenager in a crop-duster or a gun-toting neighbor with an empty stomach.

Like McDonald's, for example. Every minute, an American dies from heart disease, Goodman told us, his eyes burning a hole in the Chicken McNuggets I was dipping into hot mustard sauce as a midclass snack. Though I'd spent the last eight years worrying about the outer world, I'd never once thought about the inner world. I'd never considered that eating fried chicken and hamburgers and french fries was the culinary equivalent of walking through Baghdad at night.

Perhaps survivalism, then, is about more than shooting guns and drinking out of toilet tanks. It's also about physical exercise and a healthy diet. After all, these are more likely to keep me alive than a solar still. Even the survivalist pioneer Mel Tappan, despite all his preparations and heavy artillery, died at age forty-seven of congestive heart failure.

After Goodman led a class on heart disease, I decided to look into what I was up against. I wanted to know exactly what factors posed a threat to my life, and what their chances were of occuring, so I could better prepare myself for them.

According to the most recent census figures, 2,448,017 United States citizens died in 2005. Of these deaths, 652,091 were due to heart disease, 559,312 were from cancer, 143,579 were from a stroke, and 130,933 were from chronic lower respiratory diseases.

The fifth-largest cause of death was the one Katie and I, along with every other negativity avoider on the planet, worried about most: 117,809 Americans died from what the government calls unintentional injuries and EMTs call trauma.

This means that every year, roughly one in every 2,500 Americans is at risk of losing his or her life due to some sort of accident or injury.

According to the National Center for Injury Prevention and Control, nearly 40 percent of those deaths—approximately 45,343—were due to motor vehicles. Being in or around them is probably the most risky behavior we engage in as human beings. Evolution never designed us to withstand impacts faster than the top speed at which we run.

The next-biggest cause of accidental death, surprisingly, was poisoning, which took the lives of 23,618 Americans. This was followed by falling, which claimed 19,656 lives.

Despite the arguments of those both for and against gun control laws, Americans with firearms actually posed more of a danger to themselves than to other people: 12,352 people were murdered by guns, while 17,002 took their own lives with one. Overall, there were 18,124 homicides in 2005, in comparison with 32,637 suicides.

Other causes of traumatic death included burns and fire (6,496 people), suffocation (5,900 people), and drowning (3,582 people). Even though eight hurricanes and tropical storms (including Katrina) and seven earthquakes hit the United States in 2005, the number of people who died due to environmental disasters was a comparatively low 2,462.

Coming in near the bottom, just above overexertion (responsible for eleven deaths), was terrorism. Fifty-six Americans died of terrorism in 2005 (not counting military personnel on active duty), and none of those attacks took place on American soil.

So who was the bigger idiot: me, who had spent the last seven

years obsessed with terrorism, economic collapse, and natural disasters, or Katie, who was too scared to drive?

I'd teased her about her fears on an almost daily basis, but she had been right the whole time: the mundane is far more terrifying than the apocalyptic.

Fortunately, thanks to my EMT training, I was now prepared to handle both.

# PROPER CARE AND HANDLING OF HAWAIIAN TROPIC GIRLS

*M*y *dick, size of a pumpkin."* The ambulance stereo blasted the Mickey Avalon song through open windows. *"Your dick look like Macaulay Culkin."*

Francisco and Rob sat in the front in uniform, shouting along to the lyrics. "If you work with us, you gotta sing," they yelled back to me. It was my second day running 911 calls as a ride-along for my EMT certification.

*"My dick, bench-pressed three fifty,"* Francisco and Rob were shouting out the window, now at a blonde in a black Mustang.

Rob flipped on the siren to impress her.

The stress exposure Alwood had promised hadn't exactly ma-terialized. So far we'd been to a retirement community, where an elderly woman was experiencing shortness of breath. We'd been to a bus stop, where a plus-sized twelve-year-old had wiped out on a skateboard. And we'd been to the home of a schizophrenic teenager in a Def Leppard jersey who said she felt like someone invisible was choking her.

Though I'd practiced my new skill set by controlling bleeding, administering oxygen, strapping on a spinal collar, and attaching EKG leads, the experience exposed me more to human nature and suffering than to stress and adrenaline.

"What's your dream call?" Francisco turned and asked as we drove away from the hospital after dropping off an elderly man who'd passed out in his bathroom and defecated all over himself.

"It would be amazing to deliver a baby," I said.

"I've done that a couple of times. But my dream call is a bus full of Hawaiian Tropic girls crashes, and I have to triage all of them."

On most of the remaining calls that shift, we learned that the Rolling Stones were right when they sang, "What a drag it is getting old." Thanks to the miracle of modern medicine, the suffering, humiliation, and loneliness of a human being can now be extended for years.

And perhaps that's how it's supposed to be. In my EMT class on delivering babies, as if I needed any more evidence for my Fliesian beliefs, I'd learned that one way to make sure a newborn is healthy is to see if it's grimacing. If we come into this world smiling, something is wrong with us. So we might as well leave life just as miserable as we enter it.

This may sound cynical, but all the talk I'd heard of apocalypse, war, murder, cannibalism, and genocide over the last few years was starting to get to me. It's hard to be optimistic when you know you're going to die.

Though I enjoyed speeding through the city's backsides to find people who needed our help, all I really learned from the experience was that living to the end of my life span—77.8 years for the average American—will be a lot more pleasant if I'm surrounded by people who love me than if I'm alone. So, in addition to fitness and health, I'd have to add family to my survival stockpile. Though, unlike my other supplies, it probably wouldn't be a good idea to have one set in Los Angeles and another in St. Kitts.

On the way to class one night, I saw a motorcycle lying on the shoulder of the highway with a man slumped next to it. Every car blew past, paying him no attention. I pulled onto the shoulder, called 911, unzipped my bug-out bag, grabbed the emergency first-aid kit, and raced to his side. He wasn't badly hurt, so I pressed a two-by-four-inch piece of gauze against his arm to stop the bleeding, then secured the gauze with a roller bandage while waiting for the paramedics.

Something in me was beginning to change. I'd never stopped to help a stranger before. I'd always assumed someone else would do it—and better than I could.

Despite this, I still lacked the stress inoculation I needed. Maybe I'd just chosen the wrong ambulance crew to ride with. So, before our final EMT skills exam, I asked the guy who sat in front of me how to join his search-and-rescue team. Maybe with them I'd get the experience I needed.

Later that week, I finally received the e-mail I'd been waiting over a year and a half for. "Your application for citizenship has been approved," Maxwell wrote, "and we are now submitting your title document to the ministry of finance in order to obtain your citizenship certificate and subsequently your passport."

I couldn't believe it had finally happened. Tears of relief rushed to my eyes. If he was to be believed, and this wasn't just another stalling tactic, the backup plan I'd set in motion well over a year ago was nearly complete. I was almost a citizen of St. Kitts.

# LESSON 63

# MURDER SCENE MANNERS

f any police officer ever gives you flak, get his name, then come to me." Detective Mike Fesperman, head of the Devonshire police homicide division, was speaking in the roll-call room. It was my first time inside a police station. I always imagined I'd be entering in handcuffs, not as a potential reserve officer. "As far as I'm concerned, when you're out there, you have the same status and respect as any police officer working the scene. I have tremendous respect for what your team does."

The team was the search-and-rescue unit I'd asked my EMT classmate about, California Emergency Mobile Patrol (or C.E.M.P.), a forty-six-year-old non-profit on call around the clock for the Los Angeles Police Department and the Los Angeles Fire Department. As instructed by the team's applicant coordinator, I was wearing a white button-down shirt, black pants, and black boots. At the end of the meeting, I had to present my case to the group and then either be accepted or rejected as an applicant.

Fesperman was telling us about a body he'd found the previous week. It belonged to an artist who'd toured Europe with his work but lived as a vagrant in the park because he didn't like confined spaces. A few days earlier, he'd bought beer for some teenagers. An argument ensued, he called one of their girlfriends a bitch, and the guys returned later and stabbed him.

It was one of the most violent stabbings Fesperman said he'd seen. To determine the date of death, a maggot was removed from the body and sent to the coroner's office, where an insect expert pinpointed the murder time based on the life cycle of the larvae.

Fesperman wiped a hand across his brow. He was thin and bald, with taut skin and small, serious eyes that had seen the extremes of man's inhumanity. He'd been a homicide detective for over a quarter century, as had his father.

If learning wilderness survival had thickened my skin, joining C.E.M.P. would certainly harden what lay beneath. Beyond the exposure to stress and violence, there were survival advantages I hadn't anticipated: dozens of hours of rescue and police training; relationships with fire chiefs, police captains, and park rangers; field trips to the 911 call center, the coroner's office, and the rest of the city's emergency nerve center; police-blue uniforms, shiny metal badges, police radios, and cars with flashing lights and sirens that would get me past official roadblocks much better than my CERT uniform; and access to the government's Wireless Priority Service program, which would give my cell phone precedence over other calls during an emergency when the network was tied up. Even Spencer couldn't buy all that.

Three nights before, C.E.M.P. had helped Fesperman secure a crime scene. In this case, a fifteen-year-old girl had told her eighteen-year-old brother she'd found a rifle buried in the park. She led him to the spot and showed him where to dig for it.

But there was no gun. Unbeknownst to him, he was digging his own grave. When it was deep enough, his sister began stabbing him in the back. "What are you doing?" he cried out. "I'm your brother. I love you." Suddenly she had a change of heart, stopped, and called 911. He was brought to the hospital in critical condition.

When asked why she'd done it, she simply told the police it was time for him to die.

"Human beings are animals," Fesperman told us, scanning the room. "They're vicious animals. Some of the things I've seen them do I could never even have imagined before I started this job."

I thought my outlook on human nature was dark when I began learning survival. But the people I'd met along the way were far more Fliesian than I was—and they were in surprisingly good company. No less an authority on human nature than Sigmund Freud believed something similar. "I have found little that is 'good' about human beings on the whole," Freud once confessed in an interview. "In my experience, most of them are trash, no matter whether they publicly subscribe to this or that ethical doctrine or none at all."

Fesperman spent the next half-hour teaching newer members of the group how to preserve a murder scene. He told us not to touch the victims. Not to cover them with a blanket or sheet. Not to outline their bodies in chalk, despite what we'd seen in movies. Not to smoke, spit, or chew tobacco at the scene. And not to let anyone touch cigarette butts, bloodstains, or drag marks. All these things will contaminate DNA evidence and give the defense lawyer an opening to create doubt in the minds of jurors.

These rules, Fesperman continued, were especially important since, in a few months, the State of California planned to begin taking DNA samples from every person arrested within its borders.

I looked around the room after Fesperman said this, but no one else seemed disturbed. The official reason for inserting radio-frequency identification chips in passports and putting cameras on street corners and creating databases of people's

DNA, facial features, fingerprints, and irises is to aid investigators and to reduce crime and terrorism. But these measures come with an uncomfortable potential for abuse. At the press of a button, a government could turn these surveillance tools against its own citizens in ways that would make fascist regimes of the past seem permissive in comparison. Anonymity is a dying art.

"I look at every murder the same way," Fesperman was concluding. "Whether it's a celebrity or a vagrant, they're still someone's son or daughter."

When the meeting ended, I was interviewed by the team about why I wanted to be a member and the skills I had. After I told them about everything I'd been learning and the hands-on disaster experience I was seeking, they went silent.

"Am I in?" I asked.

"We'll let you know."

Their words traveled into my ear, formed a lump in my throat, and stopped in a knot in my chest. I had no idea if I was the kind of person they were looking for. And I wanted, more than anything, to be part of the team. Not just for the experience, uniform, and siren. I wanted to belong.

Before entering the roll-call room of the Devonshire police station that day, I was alone and adrift in a world of panicked people. But in that room, there was a community of skilled men and women with a sense of purpose and mission. Unlike the paranoid PTs and the stockpiling survivalists, these were people who weren't just trying to save their own lives. They had the resources, they understood the system, they knew the authorities, and they possessed not just the skills but the heart to help others survive in the cruel world Fesperman had described.

I walked to the police officer's break room to await the team's decision. Above the snack machines, *World's Wildest Police Videos* flickered on a small television set.

A few minutes later, the applicant coordinator, SR77—no one here used names, only numbers—appeared in the doorway. "Welcome to the team," she said. "I don't know why they just didn't tell you on the spot instead of making you suffer like that."

I broke into a wide, relieved smile. Up until now, I'd sponged knowledge wherever I could find it. This was the first time I'd felt like part of something.

When I returned to the roll-call room, I was briefed on my responsibilities, a long list of team rules, and the minimum hours I was expected to put in every month. "In an emergency like an earthquake, your first priority is the safety of your family," SR33, the team's medical officer, sternly informed me. "When that's taken care of, your next priority is the safety of your neighborhood. After that, your next priority is C.E.M.P."

I didn't realize it until that moment, but this wasn't just a survival training decision. It was a life decision. As of that evening, I was no longer Neil to these people. I was an alphanumeric sequence. I was SR14a:

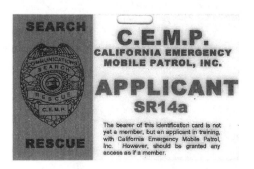

But rather than feeling like I'd lost my individuality, I felt like I'd gained a network.

# HOW TO LISTEN TO THE EARTH'S CRUST

I t wasn't long after joining C.E.M.P. that I experienced my first natural disaster.

Since being accepted as an applicant, which entailed six months of training followed by a three-month probationary period, I'd bought my first ham radio. WTSHTF, if both land and mobile lines went completely dead, I'd need some way to connect to the outside world to find out what was going on, what areas were safe, and, if necessary, call for help.

Katie didn't like the radio, especially when I left it on the earthquake channel. The station was completely silent because it was hooked up to a machine that detected seismic activity. If it started emitting tones, this meant the machine was picking up tremors and anyone listening should immediately duck, cover, and hold.

"It's like another presence in the room I can't see," Katie complained. "What if we're sleeping at night and this static comes on, and all of a sudden we hear a creepy voice saying 'I'm watching you'?"

I laughed. I actually thought she was kidding. My mistake.

"You probably think it's totally unrealistic. But, honey, it could totally happen. It's scary."

"Have you ever realized how much your life has suffered because of your fears?" I needed to do something to help Katie. If she continued like this, she'd end up as a crazy old spinster sealed in a house full of cats. Except she wouldn't even have cats to keep her company, because she'd be afraid they'd sit on her chest and steal her breath while she slept. "You're always fighting with your sister because she resents driving you everywhere, and you're constantly canceling plans and job interviews because you can't drive anywhere by yourself."

"I guess you're right about that." She took off her reading glasses and put down *The Fear Book* by Cheri Huber, which I'd bought in an attempt to help her. "I would take a taxi, but I don't trust taxi drivers." The book clearly wasn't working.

"Or you could take driving lessons."

"I don't know. I'm scared I'm going to turn the wheels the wrong way and hit things. And I'm worried about other drivers crashing into me."

"That's why you need to learn about a car before driving it. In order to engage in life fully, we sometimes have to subject ourselves to small, calculated risks. And though we can't control anyone else's behavior, we can learn to control ours to minimize those risks."

I couldn't believe the words coming out of my mouth. Somehow, during all this, I'd actually stopped acting like a whiny, clinging child and grown up.

When it came to driving lessons, Katie resisted for a couple of days. But when her sister forgot to pick her up one afternoon and she missed an interview for a job she wanted at a television production company, she relented and agreed to face her fear.

The following afternoon, an elderly Hispanic man with a white beard picked her up for her first driving lesson. Katie peered into the car, then turned to me. "I trust him," she said. "He reminds me of Lola."

Lola was our goat. She'd arrived earlier that week—affectionate, curious, and very pregnant. Katie had named her Lola after the showtune lyric, "Whatever Lola wants, Lola gets."

"She's a pretty little goat, and we will spoil her like we never spoiled Bettie," she explained.

While Katie wreaked havoc on the streets of Los Angeles, I drove to Santa Ana and took my FCC Technician Class exam, which enabled me to receive this license, a call sign, and authorization to broadcast on ham radio:

| Call Sign / Number | Grant Date | Expiration Date |
|---|---|---|
| KI6SJC | 07-28-2008 | 07-28-2018 |

| Operator Privileges | Station Privileges |
|---|---|
| Technician | PRIMARY |

STRAUSS, NEIL
8491 SUNSET BLVD 364
WEST HOLLYWOOD, CA 90069

AMATEUR RADIO LICENSE
FCC Registration Number (FRN): 0017987942
FCC 660 - May 2007

The day after my license was granted, I was home alone with the radio on the earthquake channel while Katie was taking another driving class. Suddenly, a warbling pitch emerged from the speaker.

I dove under a desk in the room and gripped its sides. As I did, the house began to shake as if a giant hand had taken hold of it and was rocking it more solidly into its foundation. Then it stopped. Its magnitude was 5.4 on the Richter scale.

No one was hurt. Nothing fell off the shelves. But it was humbling nonetheless. I thought of the analogy I'd heard so much from environmentalists: that we were fleas on the back of the earth, which was itching to get us off.

Since I was safe, I called Katie to make sure she was all right. Then I walked outside to check on Lola and ensure the neighborhood was secure. Next I called C.E.M.P. and asked if they needed me.

It felt good to know what to do.

# WHERE TO SWIM
# ACROSS THE BORDER

It was a minor miracle: Katie was actually driving me to the UPS Store, where I needed to get fingerprinted for my EMT license. She was self-conscious behind the wheel, hesitant about controlling an object much heavier and more powerful than her, but she had done it. On her third try, she'd passed her driving test and received a license.

"I feel like my entire life has changed," she gushed. In the last week she'd re-enrolled in school and driven herself to half a dozen job interviews. "Now, instead of having to depend on everyone, I can depend on myself. It's like freedom." She paused to make a wide, sloppy turn onto Ventura Boulevard, then continued. "It's weird, though. When someone wants me to meet them somewhere, my first reaction is to get stressed out and think, Fuck, I can't."

"So what do you do to get over that?"

"I just get in the car and drive," she said, as if the solution had been that simple all along.

"But why are you able to do that now when you couldn't before?"

"Because now that I've had more practice driving, I'm confident," she answered as she careened over a speed bump and landed with a thump in the parking lot. "I trust myself now."

Funny, I thought—that's exactly how I felt after practicing the survival skills I'd learned. Though, hopefully, I was a better survivalist than she was a driver.

At the UPS Store, I watched as a college student wearing white gloves rolled my fingers, one by one, onto the scanner of a computer. I watched as my fingerprints appeared, one by one, on the screen. And I watched as he clicked on the submit button. Within seconds, my fingerprints were on file with the United States Department of Justice forever. They now had identifying information I couldn't simply change with a new document.

And all in exchange for this:

County of Los Angeles
Emergency Medical Services Agency
EMERGENCY MEDICAL TECHNICIAN-I

**Neil Strauss**

has fulfilled the requirements for certification in California

Certification#   53773
Cert. Date       9/2/2008
Exp. Date        9/30/2010

*William J. Koenig*, MD.
William J. Koenig, M.D. Medical Director

"You're on their records now," the student said afterward. Even he knew I'd just done something irreversible.

I guess I'd made a choice that afternoon.

I'd spent days debating whether to take this step or not. On one hand, it meant sacrificing the privacy I valued so much. But without an EMT license, I wouldn't be able to treat patients for C.E.M.P. or work for local ambulance companies. Ultimately, I decided it seemed wrong not to get the license just because of the slim chance that, in some Orwellian future, the government might come after me. In Kurt Saxon's words, "Paranoia doesn't pay."

For perhaps the first time, my fight instinct had beaten my flight instinct.

Besides, I now had the skills to sneak across the border anyway. I'd even talked to my new law enforcement friends and learned that the best place to make my getaway was across the Rio Grande near McAllen, Texas, which currently wasn't patrolled by boat between midnight and eight A.M.

Paranoia may not pay, but it does sometimes require payment. And that payment comes in the form of reassurance.

As for the privacy I had to sacrifice, like Katie overcoming her driving phobia, I learned that doing what I feared didn't make my concerns any less valid. But it did destroy the immobilizing power they held over me. So, since my prints were now on file, there was no longer any harm in applying for the concealed-weapons permit I'd studied for at Gunsite:

Or taking a shooting test to get a permit to carry a firearm openly in California:

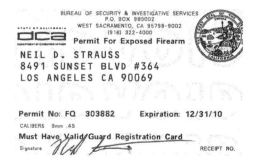

Or taking a written test to get licensed as a security guard:

BUREAU OF SECURITY & INVESTIGATIVE SERVICES
P.O. BOX 989002
WEST SACRAMENTO, CA 95798-9002
(916) 322-4000
**Guard Registration**

NEIL D. STRAUSS
8491 SUNSET BLVD #364
LOS ANGELES CA 90069

Registration: G 1622603 Expiration: 12/31/10

Additional Permit Required to Carry Firearm
Signature RECEIPT NO.

If anyone was in trouble, they were definitely going to want me around now. I was licensed to survive.

"When I was a teenager, I used to think I'd rather die than grow old." I told Katie as she drove home from the UPS Store. "One of my heroes was a poet who pledged to kill himself when he turned thirty. Now I'm obsessed with living as long as possible. I don't know what happened to me."

"Maybe," Katie said as she swerved onto Laurel Canyon Boulevard, narrowly missing a car making a right-hand turn, "you just needed enough experience to show you that life is too beautiful to want it to end."

# HOW TO LIVE LONGER

**1. SLEEP BETWEEN SIX AND SEVEN HOURS A NIGHT.**

**2. DON'T SMOKE.**

HAZARDOUS MATERIALS **KEEP OUT**

**3. GET A PET.**

**4. SEE A DOCTOR FOR AN ADVANCED CHOLESTEROL TEST AND ALTER YOUR DIET ACCORDINGLY.**

**5. GET CLOSER TO FAMILY MEMBERS AND LOVED ONES.**

**6. HAVE AN ACTIVE SEX LIFE.**

7. BE OPTIMISTIC.

IS THAT THE FORMULA FOR THE ANTIDOTE?

8. REDUCE STRESS AND AVOID GETTING ANGRY.

NOT EXACTLY, BUT IT WILL ALLOW US TO CONTINUE OUR RESEARCH WITH NEW VIGOR.

9. HAVE A PURPOSE IN LIFE AND TAKE ON NEW CHALLENGES.

WE HAVE ANOTHER TOP-SECRET MISSION FOR YOU IF YOU'RE WILLING TO ACCEPT IT.

10. REDUCE YOUR CALORIC INTAKE, AVOID PROCESSED FOOD, EAT MORE NUTS, AND CONSUME FRUITS, VEGETABLES, RED WINE (MODERATELY), AND OTHER SOURCES OF ANTIOXIDANTS DAILY.

WE'LL NEED A CASE OF WINE AND A KEG OF BEER TO, UM, COMPLETE THIS FORMULA AND STOP THIS PLAGUE.

11. PERFORM DAILY PHYSICAL ACTIVITIES YOU ENJOY.

YOU GOT IT.

ONE WEEK LATER...

12. SURROUND YOURSELF WITH PEOPLE WHO PRACTICE THE PREVIOUS ELEVEN STEPS.

HELLO, IS *ANYONE* HERE?

END.

## LESSON 66

# SECRETS FOR CIVILIZING LONE WOLVES

The call came at 4:48 P.M.

"There's been a train accident," SR77 said grimly. "We're staging at Rinaldi and Canoga in Chatsworth. I don't have any further information right now."

"I'll be there in twenty minutes."

I slipped into my uniform, pinned on my badge, loaded my tactical belt, grabbed my EMT jump kit, and raced to the car. As I neared the staging location, four policemen stood in front of yellow caution tape stretched across Canoga Avenue. I showed my C.E.M.P. badge and they let me through the roadblock, no questions asked.

We'd already been involved in a number of incidents—most recently, helping to capture a serial arsonist in Griffith Park. But this was the biggest disaster we'd been activated for since I'd joined. In fact, it was one of the biggest disasters in Los Angeles since I'd moved there nine years earlier. In learning to run away from catastrophes, I'd somehow ended up running to them.

A Metrolink commuter train carrying 222 passengers had collided headfirst with a Union Pacific freight train. Near the scene, I found the rest of the team amid a sea of police cars and fire trucks. Smoke billowed into the air, while stray passengers wandered dazed and bloody in the street. Nearby, firemen waited impatiently for permission to cut through dead bodies to get to victims trapped beneath. Though the news had reported only three casualties so far, it was clear there were many more.

Though I wanted to run to the train and help, this was the incident command system in action and we had to wait for our assignment. In the system, either everyone is a hero or no one is. Individual initiative costs lives.

"Who here's an EMT?" asked SR07, our incident commander. He'd just received his instructions on the radio.

I raised my fortunately fingerprinted hand, as did five other members of the team. "I need you to grab your medical bags and go to Chatsworth High School. We'll be staging there."

We jumped wordlessly into my car and drove to the school, where we turned the auditorium into a reunification center for families and the gym into a treatment center. In the meantime, rescue workers from the Red Cross, the Office of Emergency Services, the police department, and the fire department began arriving.

While the severely injured victims were transported to area hospitals, the walking wounded were sent to us. They trickled in with cuts, bruises, scrapes, and sprains. Around their necks hung the triage tags I'd learned about in CERT and EMT classes. I never thought I'd see them used in a real disaster so soon.

As we worked, a van belonging to a computer repair company pulled up in front of the gym to distribute pallets of water. Local residents came by with boxes of donuts and thermoses of coffee. Pizza stands, supermarkets, and drugstores contributed food and supplies. The generosity from the community was staggering. It flew in the face of my Fliesian beliefs.

My first patient was a bald Hispanic man in a large button-down shirt and khaki pants. He limped toward my station and collapsed into the chair. After quickly checking his ABCs—airway, breathing, circulation—I knelt down and inspected his legs. As I did, I was overwhelmed by a sense of purpose. Prior to this moment, my life had been dominated by the pursuit of pleasure, personal growth, and survival. I'd never imagined I'd be doing something that was actually helpful to others, or that I'd find it so fulfilling.

"Me and some of the passengers wanted to go back in for the others, but the car was on fire," he said as I checked for fractures and deformities. He seemed to be replaying the scene in his head, trying to figure out if there was anything he could have done to help his fellow commuters. "But we got scared. It was too dangerous. We tried, though. We really did."

Unlike what the survivalists, the PTs, *Lord of the Flies* and Sigmund Freud had led me to believe, it seemed that tragedy also had the power to bring out the best in people. Not just the firefighters and police officers who worked around the clock. Not just the locals who arrived en masse to volunteer. Not just the businesses that brought truckloads of supplies for neither monetary nor marketing gain. But even the victims themselves tried their best to help one another.

They might have behaved differently if their lives were still in danger, resources were scarce, and they had to compete to survive. But once they were safe, it seemed that people's first instinct was to look after one another and support their community. Maybe I needed to consider modifying my Fliesian philosophy. If people were animals, then like most animals, they were essentially harmless most of the time—unless they felt threatened. That's when they became vicious.

As we treated more survivors, cookies, granola bars, hamburgers, pizza, and energy drinks continued to arrive by the armful. When I thanked one woman for her generosity, she re-

plied, with a smile, "It's not looting if you leave a note." I looked up and recognized her from my CERT class. I guess I was no longer a lone wolf.

When the trickle of survivors stopped, we were sent to the crash site to join the rest of the team. The riot of toppled, crushed, derailed, and accordioned train cars was one of the most brutal things I'd ever seen. The Metrolink locomotive had been pushed back into the first passenger car, shearing through the metal to come to a stop more than midway into the carriage. The force of the impact had knocked the car off the tracks and onto its side, where it lay decimated. It seemed doubtful anyone in the front of the carriage had survived.

Our task was to help light the scene so firemen could extricate the remaining bodies like this:

On the ground outside the decimated car, a lone sneaker lay in a pool of diesel fuel. I wondered about the fate of its owner. Meanwhile, paramedics taped a yellow tarp over the back windows of an ambulance, preparing it to receive the body of a female police officer who'd been a passenger in the front car.

I walked back to the second car, where I noticed a pile of

bloody rags in the doorway. Beyond them a man lay sprawled in the stairwell, partially covered with a sheet. It was the first dead body I'd seen in my life. His right hand was extended. Next to it, a cell phone lay flipped open, as if he'd been making a call, unaware it would be the last thing he ever did.

It was my worst fear come to life. The torn metal, the shattered glass, the bodies ripped from life: this was what the world really looked like through apocalypse eyes.

And it woke me up to something I'd spent the last three years trying to fight: my own powerlessness.

In that moment, it felt like there was no other being out there or up there or anywhere who cared about us. We're just fragile machines programmed with a false sense of our own importance. And every now and then the universe sends a reminder that we don't really matter to it, hurtling us into confusion and a panic for answers that will allow us to resume our program again.

On the other side of the train, a blood-soaked white blanket lay over part of the body of the engineer. It looked as if he'd tried to jump out of the train at the last minute, only to be crushed by the engine as it toppled onto him. If he'd made it just three feet further, he would have lived. Perhaps survival, like success, is just a matter of talent, luck, and timing; maybe that would be a more accurate motto for the Survivalist Boards than "endure—adapt—overcome."

Further back, up the side of the embankment, a large, lumpy yellow tarp lay spread across the dirt. As if to eliminate any doubt as to what it concealed, a cell phone began ringing from underneath.

Now I knew how I responded when plunged into the fire. I didn't get physically sick. I got spiritually sick.

A few minutes later, a team from the coroner's office arrived and began pulling back parts of the tarp. Bodies lay underneath in various stages of deformity. One man was eviscerated, his organs spilling onto the ground. A young man's leg was amputated beneath the knee, his face crushed beyond recognition. And a woman's bones were broken in so many places she looked like a discarded marionette.

They were all on their way somewhere. They all had unfinished business to take care of. And it didn't matter whether they were the good guys or the bad guys, honor students or struggling businessmen, faithful wives or cheating husbands, hard-core survivalists or frail snivelers. They were all dead, indiscriminately murdered by fate, if one can even believe in such a word after an accident like this.

I followed the path of our floodlights further along the embankment to the side of the freight train. On the ground alongside it, I noticed a thick pool of red. It appeared to be the biggest bloodbath I'd seen yet—until I noticed blue vats nearby and realized, thankfully, they were all filled with strawberry preserves, which must have been part of the train's cargo.

When I left the scene and returned to Chatsworth High, there were still families sitting outside the reunification center. They were waiting for their sons, their daughters, their fathers, their mothers to return. But I had just seen the loved ones they were waiting for under yellow tarps and white sheets. My eyes misted with my first tears of the day when I saw the hope in their faces. There was no more hope.

Twenty-five people died in the crash that day and 134 were injured. It was the worst train disaster in California history. "I've been in the force for twenty-one years and I've never seen anything like this," the captain of the Devonshire police told us during a combination stress-management session and award ceremony afterward. "I want to thank you for all you did. If there wasn't a response like this, more people would have died."

That night, I noticed that the change that had begun in EMT class had taken root. I no longer wanted to just get a C.E.M.P. uniform and a little stress inoculation, and then run. I may have felt that way when I was taking the fire department's CERT course. But now, as I lay in bed, instead of imagining how to escape the city and talk my way past barricades, I thought about what I would have done if I was the first person on the scene of the train crash and had to triage the victims until ambulances arrived. Whenever I heard the screech of car brakes outside my house, I tensed, ready to grab my EMT jump kit and run to the street if the sound of crushing steel and breaking glass followed.

My first instinct was no longer to flee in the face of disaster. Instead, if that disaster posed no threat to me, my first instinct was to stick around and try to be of service.

Somehow, since meeting Leonard Cohen, I'd accidentally

completed my transformation from runner to fighter. And perhaps that made me a better survivalist, because my preparations were no longer just for my own continued existence but for that of my community.

After all, for those who look the awful truth in the eye and have the courage not to turn away, there's only one way to carry on with life: be good to and look after each other. Because if we just give up, Leonard Cohen's fear comes true and the spirit dies.

## LESSON 67

# HOW TO TELL WHICH SIDE GOD IS ON

I got my St. Kitts passport." Spencer had flown into town, probably just to gloat.

"I guess this means you don't need to buy your own island anymore," I replied, though I secretly hoped he'd still get a submarine so I could ride in it. "How'd you get yours before I got mine?"

It was two weeks after the Metrolink crash, and we were eating Mexican food on Sunset Boulevard. Spencer's Hamptons roommate Howard had recently fled the country after the mortgage company he helmed had collapsed. And Adam, the venture capitalist, was living in Austria. We were the last ones left.

"The developer came back to us with a modified agreement, so we were able to buy the house." He bit into a burrito the size of a baby's arm. Four women rose from the table next to us, leaving their food half-finished. "And not a moment too soon."

The day before, the government had seized Washington Mutual Bank. And in previous weeks, some of the country's biggest mortgage, insurance, and brokerage firms had either gone bankrupt or been bailed out from the brink by the government, including Lehman Brothers, Fannie Mae, Freddie Mac, and AIG. On top of everything, Tarasov and Associates, the company Spencer and I had used to protect our assets, was under investigation by the IRS. I hoped they'd remembered to protect their own assets.

Everything Paul Kennedy had predicted in *The Rise and Fall of the Great Powers* seemed to be coming true.

I was reminded of a story I'd once heard about two men wrestling. A spectator asks a wise man, "Whose side is God on?" The wise man replies, "God is on the side of the winner."

"We lost an empire in eight years," Spencer continued. Around us, billboards several stories tall advertised iPods, Calvin Klein bras, *Seinfeld* reruns, and, with the slogan "Drive like there is a tomorrow," the Mini Cooper.

"What I want to know is how that will trickle down to the average person." I wondered what future history books would say about us. Between the nuclear bombs we'd dropped in Japan, the massacres in Vietnam, the invasion of Iraq, and the attempts to politically and economically enslave less developed countries, it seemed likely that we'd go down in history as the bad guys. "Look at Mao's Great Leap Forward. As a consequence of bad policies, more than twenty million Chinese citizens died of starvation."

"There are a lot of doomsday scenarios, but I don't think we're there yet," Spencer replied. "What we're seeing is the painful shock of the U.S. unwinding its heavy leverage." The traffic on Sunset Boulevard flew by. Wide black Hummers, boxy pink Mini Coopers, squat yellow Porsches. You'd never know from the surface that the country was sick. "The hope is that the world has a motivation not to let that happen, because we're their customer and they have a vested stake. Imagine losing twenty-five percent of your business. If America goes down, China and Russia and everyone else are going to be pulled down as well. And that's going to lead to a lot of turbulence and revolution."

"So are you prepared for a world crisis?"

"I feel like I am." Two men in button-down shirts with crosses and skulls on the back walked past, both talking on their cell phones. As long as we have cell phone reception, as long as we're able to connect to the Internet, as long as we can turn on the TV and see the same familiar faces, then we'll never truly believe anything is wrong. By the time we lose reception, it'll be too late. "We've done what no one else took the time to do. If things get to the point where it's really bad for us, it's an easy plane ride to safety. We may have to lose some money and rebuild somewhat, but we'll be fine."

"I'm calling Maxwell as soon as I get home," I pledged as we left.

Two weeks later, Maxwell e-mailed to let me know that my St. Kitts passport was waiting for me in his office. Finally, my backup plan was complete. I leapt up from the computer and told Katie the news.

"What if he's lying?" Katie responded with her usual quick thinking. "We should go there and find out. St. Kitts is a small island—we can hunt him down and interrogate him and get your money back."

"How will we do that?"

"We'll hide around the corner from his office until he gets there, then jump out and tell him he has to give you your money back or else."

"Or else what?"

"Or else we go to the police. And if they don't believe us, then we bribe them. Money talks, babe."

Katie was right, to a point. I needed to get back to the island as soon as possible to see if Maxwell had really come through.

I thought about how much had changed since I'd first walked into his office. Back then, I was scared. Now, though the world had become an even scarier place, I wasn't motivated by fear. If there was a disaster in America, I wanted to be here for it. I

wanted to stay and look after my neighbors and do recovery work with C.E.M.P.

Yet at the same time, I missed St. Kitts. I liked the people, the beaches, the climate, the cleanliness, the rum punch. It was the closest I'd been to paradise, and the few times I'd visited were happy and productive. Unlike America, which had let me down over the last eight years, St. Kitts hadn't disappointed me yet. It was a new relationship. It was the Wild West with an ocean view.

Before flying back to St. Kitts, I called Wendell Lawrence, who'd originally urged me to move there. I told him about the rescue work I'd been doing and asked if there was any way I could be of help to the island. He put me in touch with the National Emergency Management Agency, the St. Kitts equivalent of FEMA.

And so, after waiting for what would be the last time in the St. Kitts tourist lane to clear immigration, I returned to the Caribbean—not just to take advantage of its citizenship program, but to be useful.

## LESSON 68

# THINGS TO DO WHEN YOU'RE A DUAL CITIZEN

As soon as I touched Kittitian soil, I took a taxi straight to Market Street. So much time had passed since I'd last seen Maxwell that I made the mistake of trying to engage him in small talk. "Any luck working less and golfing more?"

"No." The word seemed laden by the weight of the world. "Unlike you, I have to work all day—not just sunbathe and play tennis."

I don't even play tennis.

Maxwell started filling out the last of my paperwork. Every laborious action seemed like a message letting me know that he'd rather be golfing. But this time I didn't care. It didn't matter if he was happy, sad, friendly, mean, or illiterate. He'd done his job. And though he'd done it slowly, perhaps that was just the pace of life down here.

As he handed me a manila envelope, my heart filled with the warm, expansive feeling that Greg from the Sovereign Society told me he'd felt after leaving New Zealand. It was as if a large white building had been lifted off my chest.

I thought of Tomas and the 4,000 immigrants receiving their American citizenship. This was my ceremony. Though it was slightly less august, it was no less poignant. I shook Maxwell's hand, thanked him, and then, in a careless burst of optimism, invited him to dinner to celebrate.

"I'm exhausted," was his reply.

I left the office clutching the manila envelope as if it were a newborn child I didn't want to drop. The street was dense with the sounds of soca, the smell of fresh-baked bread, the bustle of teenagers on their way to this or that street corner. I wanted to find someplace quiet to open the envelope. Somewhere it wouldn't get snatched out of my hands.

Nearby, the tight cluster of colorful buildings and cramped shops parted to reveal a cathedral. I walked down the street and entered the dark church. Aside from the attendant at the door, I was the only person there. I sat in a pew, beneath an organ with pipes as thick as my torso. And, with Jesus staring down at me from the cross, I undid the clasp on the envelope.

The first document I saw was a white piece of paper, which I pulled out to examine:

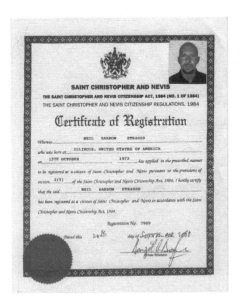

The sight of it took my breath away. Especially when, beneath the thuglike photo I'd taken over a year ago in Koreatown, I read the words CITIZEN OF SAINT CHRISTOPHER AND NEVIS.

I reached back into the envelope and grasped a thin booklet. A lump formed in my throat when I pulled it out:

I opened it, my excitement building to a climax, Jesus craning his neck behind me:

There it was. Finally. My name and mugshot on a non-U.S. passport. Even secret agents had to forge their documents. This was legitimate. As Wendell had said, my future wife and children would be Kittitians too.

As I stared at it, I realized that a second citizenship was no longer something I needed, but something I wanted. Nationality mattered a lot less to me than it had two years earlier. I had survived just fine with a knife in the forests of America. And I'd survive just fine with a knife in the jungles of St. Kitts.

Just to make sure I could get by on the island without help from the outside world, I found a local nature expert named Kris, who helped me get to know the local plants and wildlife, which consisted mostly of monkeys and, ironically, goats.

Though I worried I'd have buyer's remorse after such a long, costly journey to get such a little booklet, I was elated. I had a beautiful apartment on a beautiful beach in a beautiful country. I was an American. I was a Kittitian. The world was that much more open to me.

Besides keeping the passport as a backup in case of emergency—which didn't seem so paranoid when three weeks later, during a terrorist attack in Mumbai, eyewitnesses reported that gunmen specifically asked for people with American and British passports—I already had three immediate uses for it. First I booked a trip to Cuba. Then I called back the banks that had turned me down. They might not want to do business with Americans, but there was nothing wrong with doing business with Kittitians. And, finally, I took Spencer up on the advice he gave me when I first met him and started a publishing company, luring my first authors with an offer to work in my compound in St. Kitts.

By compound, I meant my apartment with a stockpile of supplies.

Unfortunately, the first author I signed died of heart failure before making it to the island. He did, however, live six years beyond the average American life expectancy. His name was Larry Harmon, but he was better known as Bozo the Clown, one of television's most popular children's characters.

Even with the makeup off, he was the happiest person I'd ever met. His secret to longevity, he told me one night over dinner, his face glowing with enthusiasm, was "Just keep laughing."

Along with my passport, guns, water, bug-out bag, medical supplies, a loving family, a healthy diet, and exercise, I promised him I would add laughter to my survival stockpile.

A good soldier is always prepared.

## LESSON 69

# WHAT TO DO WHEN THE LIGHTS GO OUT

Nearly five hundred Kittitians stuffed themselves into Sugar's, a local bar and pool hall, to watch the 2008 U.S. election. Among them was a small smattering of foreigners, including Katie, Spencer, and myself.

"I like it here," Katie chirped as someone handed her an Obama pin. "Everywhere you look, it's really lush and green. There are little monkeys. And the people are nice. I feel safe."

I agreed. I'd come to trust her instincts.

We had voted by mail, then flown to St. Kitts. Spencer was worried that if Barack Obama lost, there would be race riots at home. And though I didn't worry about unlikely events like that anymore—especially since I had a ham radio, a network, and experience to keep me informed and out of the danger zone—it was a good excuse for my first trip to the island as a citizen. Judging by the enthusiasm at Sugar's, where Kittitians wore Obama shirts and sang, "Hey hey hey, good-bye," every time McCain lost a state, people would probably be just as angry here also.

As CNN blared on a bass-heavy sound system more accustomed to soca and reggae music, I thought back to the election four years ago, when the idea of leaving the country first entered my mind. And I hoped that this time we would redeem ourselves in the eyes of the world—and of history.

As midnight neared and Obama's electoral votes began to

surge, bartenders handed out plastic glasses and filled them with complimentary champagne. Then, suddenly, a sound like that of a record slowing down filled the bar. The televisions shut off, the lights flickered out, and the room went silent.

Another blackout had struck.

Fortunately, this time, I was prepared.

I switched on a flashlight and found Spencer and Katie. "We can go watch the election back at my, um, compound. I have a generator."

On the way home we took a detour along the Caribbean, past the open-air beach bar where I'd first made the decision to buy the apartment that would give me citizenship. We noticed it had a generator running and CNN blasting, so we decided to watch the results over rum punch on the beach. If you're prepared and don't panic, I'd come to realize, most of life's emergencies are merely inconveniences.

When the forty-fourth president of the United States was announced at midnight and the crowd at the bar burst into applause, a wave of relief spread through the three of us. It felt as if we'd been holding our breath for eight years, waiting to exhale. It's not that George Bush was a bad guy. There are no bad guys. Just people who are bad at their jobs.

In a way, I was even grateful to him. If it weren't for his administration, I wouldn't be in a second home on a moonlit Caribbean beach, with a completely new outlook on life and my place in it. St. Slim Jim, patron saint of dual citizens and rescue workers.

I just hoped the changing of the guard in America wasn't too late. As the gaping hole at Ground Zero had been reminding us for the last seven years, it's easy to tear something down. It's difficult to resurrect it.

"There's still hope for America after all," a celebrating Kittitian in a bright red shirt proclaimed nearby. There was that danger-

ous word again: *hope*. But I felt it too—for the first time in seven years. I was reminded of the party in the White House on the night we safely entered this millennium, when we expected the best but weren't prepared for the worst. This time around, not only was I prepared, but I'd learned that hope is a passive emotion. It's the last survival skill of the powerless. In the face of the unknown, I prefer action.

That's why, when I'm not in St. Kitts, most days you'll find me in Los Angeles, doing search and rescue as SR14 with C.E.M.P., training with the Disaster Communications Service as ham radio operator KI6SJC, working local medical events as EMT number B1892201, running mass-casualty incident drills with CERT battalion 5—or milking goats in my backyard.

Those who run from death, like the survivalists in their bunkers and the permaculturists in the forest, also run from life. As an EMT, as a C.E.M.P. member, as a latecomer to the world of outdoor adventure, I'd discovered that the opposite is also true: those who run to death also run to life.

When you walk to the very edge of the abyss, and you lean over and peer as deep into the blackness as you possibly can, and maybe you even lower a hand into it and pull someone out who's not supposed to be there, that's when you feel alive.

I used to wonder if Kurt Saxon, Tom Brown, Bruce Clayton, and all the other survivalists I met ever regretted dedicating their lives to a skill set they never had to use. But now I know the answer. They use those skills every day. Because after three years of searching and learning and accumulating, I've learned that it isn't actually survival these skills bring. It's peace of mind.

I now know that I can take care of myself and my loved ones. But until the day comes when I have to do that, I'm going to be taking care of everybody else.

# EPILOGUE

She was a nineteen-year-old student. She sang, played piano, and went to church every Sunday. Today, there was a C.E.M.P. call-out to Northridge, where an SUV sped through a red light and hit her as she was crossing the street.

Her body flew several dozen feet through the air before landing face-first on the ground. The jewelry she was wearing clattered across the intersection. The artwork she was carrying scattered in the wind.

She seemed talented. She seemed smart. She seemed generous.

She never had a chance.

It could have just as easily been me. It could have just as easily been you. But it was her.

Tomorrow, though, is another day.

# THE PARTING WORDS OF THE FISHWIFE SIDUR TO GILGAMESH :

*"When the heavenly gods created human beings, they kept everlasting life for themselves and gave us death. So, Gilgamesh, accept your fate. Each day, wash your head, bathe your body, and wear clothes that are sparkling fresh. Fill your stomach with tasty food. Play, sing, dance, and be happy both day and night. Delight in the pleasures that your wife brings you, and cherish the little child who holds your hand. Make every day of your life a feast of rejoicing! This is the task that the gods have set before all human beings. This is the life you should seek, for this is the best life a mortal can hope to achieve."*

# EXTRA CREDIT

There was so much I did during the last three years to prepare myself that there wasn't room for all of it in this book. In fact, there isn't even room for complete endnotes. So, for the research-inclined, I've compiled a list of sources for the preceding facts and information. You can find it online at *www.fliesian* *.com,* where I've also posted the contents of my bug-out bag in case you want to build one yourself. If I've left anything out—of either the bibliography or the bug-out bag—email me at *stslimjim@gmail.com.*

As for the other people who contributed to my survival skill set but weren't mentioned elsewhere in the book, I'd like to single out a few of them here.

First of all, thanks to Lance Harris, who not only helped me get my bullet-proof vest, my California guard card, and my exposed weapons permit, but spent long nights talking to me about building the perfect team of people to survive an apocalypse. "Money doesn't matter," he told me the day we met. "People can store and save all they want. But when they meet a guy like me in a doomsday scenario, I'm going to take it all." And I thought, "I want to make sure this guy's on my side."

Thanks to Sid Yost of Amazing Animal Productions (and Trevor and Juliet) for helping me find such a well-dispositioned goat—and for the hands-on experience of learning to survive wild animal attacks. I'm sure I'll never have the opportunity again to feed a bear a donut, put my arm in a lion's mouth, or get my face licked by a wolf.

Thanks to Jon Young, who helped me learn nature awareness not just through his Kamana program and his bird-call expertise but through directions to his house like, "turn left at the hill covered with Monterey pine trees."

Thanks to Roger Gallo, who wrote *Escape From America* a decade ago and gave me plenty of low-budget options for escape. "I would say there's a number of Central American nations, beginning with Belize, where a person could go in and literally live off the land or the ocean," he began. "You also have, with varying degrees of social freedom, Honduras, Nicaragua, El Salvador, Guatemala, Costa Rica, and Panama. Chile is getting expensive, but there are other nations on the west coast of South America that are still reasonable: Peru, Ecuador, and Bolivia, where there are artist expat havens that many people don't know about. And in the Caribbean, the Dominican Republic is a bargain."

Then there's Jason Hadley, the first person I dared to meet from the Survivalist Boards, who kindly came over and showed me the contents of his bug-out bag, which were an inspiration, especially the pocket containing barter items like coffee, tobacco, and tampons. And my second friend from the Survivalist Boards, ace carpenter Morgan DiNavia, who taught me the credit-card knife and helped me build my hides.

Thanks to Gordon West for helping me get my Technician class license, introducing me to the magic of ham radio, and keeping the spirit of amateur broadcasting alive and thriving.

Thanks also to AJ Draven, for teaching me Krav Maga every Monday evening; David MacGregor of Sovereign Life, for a clear understanding of the nebulous world of the PTs; Nighthawk for the Tom Brown tracker knife and military consulting; those-who-shall-not-be-named in government intelligence for helping me confirm some of the information in this book; Simon Talbot and Captain Don for the boat know-how; Steve McGowan, Robert Drake, and Jim Vanek for the gyrocopter expertise; Sergeant Holcomb for the helicopter help and the advice to "think like an Indian"; Kate Murphy for being my mentor at C.E.M.P.; Sazzy Lee Varga for helping me get around goat-zoning ordinances; Mark Gallardo at Gun World for the patience; Rafik at St. Christopher Club for taking me to the killer bees; and Victor for helping me recover from them.

The thanks continue for Jeff Graham of Kern Rokon for helping me get started on the bike; Dessi Quito for the motorcycle escape; the staff at the L.A. Ranger Station for putting up with the phone calls; Jeanne O'Donnell at the L.A. County Office of Emergency Management; Marie at the Canadian Embassy; Alissa Morales at Los Angeles County Health Services; Fred Mehr in the Hazardous Materials Unit of the Governor's Office of Emergency Services; and Kevin Reeve, who provided information used in the dog-fighting, restraint-escaping, lock-picking, and car-chasing illustrations.

Reeve also taught me another important piece of survival information that I'd previously botched. "The secret to peeing in plastic water bottles," he told me after I saw him sneaking into an alley with a water bottle one evening, "is to push down with your thumb on top of your dick to create a air pocket. If you make a tight seal, you're screwed."

Above all, I'd like to thank—and definitely apologize to—Kristine Harlan (known to the people she's suspicious of simply as Agent 99) for helping to source gyrocopter pilots, gun collectors, goat breeders, radio clubs, FEMA experts, and all manner of things that will make her never want to assist with research for me again, though I hope she does.

Thanks to Cliff Dorfman, Kelly Gurwitz, and Grace Adago, who patiently listened to me read the entire book over the phone and provided a tireless stream of sage advice. For four layers of fact-checking (which I hope was enough), thanks to Ben Scott, Ray Coffey, Alex Willging, and Noelle Garcia. And thanks to my crack advance-reading team: Tim Ferriss, Jackie Dawn, Jonathan Glickman, Greg Fellows, Anna David, Ben Wyche, Katie Field, Don Diego Garcia, Dean Jackson, Stephanie Decker, Stephen Miller, Jeffrey Ressner, Thomas Scott McKenzie, Ben Rolnik, Kaia Van Zandt, Bravo, Tynan, Ghita Jones, Todd, D the R, and M the G.

Thanks to my lucubrating and sempiternal editor, Cal Morgan, and the other deadline survivalists at HarperCollins, for a long and continuing relationship, particularly Carrie Kania, Lisa Gallagher, Brittany Hamblin, Alberto Rojas, Kolt Beringer, and Kyran Cassidy. Thanks, as always, to Ira Silverberg, my trying-not-to-be-snobby-yet-growing-more-snobby-by-the-year agent, and also to Ruth Curry. To Todd and Andrea Gallopo, Jeremy DiPaolo, Danielle Marquez, Bahia Lahoud, and the Meat and Potatoes design dream team for their heroic work. To Bernard Chang, whose wrists will hopefully forgive me for this. To Anthony Bozza, for starting the publishing company with me. And to Liam Collopy, Shari Smiley, Michael Levine, Peter Micelli, Gregory McKnight, Ramin Komeilian, Pedro Gonzalez, Peter Meindertsma, Omar Anani, and Julian Chan for sundry services rendered.

Note that the sequence of a small number of events, such as the order in which some lessons were taught in the classes, has been rearranged for the sake of continuity. In addition, some of the identifying information in the documents included in this book has been changed to prevent identity theft. Not that I don't trust you, of course.

I just don't trust the people you hang out with.